Comparative Philosophy and the Philosophy of Scholarship

Comparative Philosophy and the Philosophy of Scholarship

On the Western Interpretation of Nāgārjuna

Andrew P. Tuck

New York Oxford
OXFORD UNIVERSITY PRESS
1990

Oxford University Press

Oxford New York Toronto
Delhi Bombay Calcutta Madras Karachi
Petaling Jaya Singapore Hong Kong Tokyo
Nairobi Dar es Salaam Cape Town
Melbourne Auckland

and associated companies in
Berlin Ibadan

Published by Oxford University Press, Inc.,
200 Madison Avenue, New York, NY 10016

Library of Congress Cataloging-in-Publication Data
Tuck, Andrew P.
Comparative philosophy and the philosophy of scholarship : on the Western interpretation of Nāgārjuna / Andrew P. Tuck.
P. cm. Includes bibliographical references.
ISBN 0-19-506156-X (alk. paper)
1. Nāgārjuna, 2nd cent. Mādhyamikakārikā. I. Title.
BQ2797.T83 1990
294.3'85—dc20 89-25510 CIP

2 4 6 8 9 7 5 3 1

Printed in the United States of America
on acid-free paper

PREFACE

It is a commonplace of contemporary scholarship that any theory or interpretation necessarily reflects the assumptions of its author and its readers. As the aims, conscious and unconscious, of scholars change, their readings of texts will change as well. To this extent, their readings are—sometimes positively, sometimes negatively, always productively—isogetical: they reveal far more about the views of scholars and their scholarly eras than exegesis is said to do. This volume presents a case study of the effects of changing biases on the understanding of a single, highly interpretable text.

A sequence of distinct interpretive fashions is discernible among those Western (or Western-trained Asian) scholars who, since the mid-nineteenth century, have written on Indian philosophical and religious thought. Tracing the radical discontinuities in interpretation of the second-century Indian-Buddhist text, Nāgārjuna's *Mādhyamikakārikā,* it is possible to chart three phases of interpretive style. Nineteenth-century idealists from Schopenhauer on viewed Indian thought as a response to the problem of the relation between appearance and reality and found their own concerns mirrored in Upaniṣadic, Vedāntin, and Mādhyamika writings. Accordingly, Nāgārjuna was read as if he were a Platonic or, more usually, Kantian transcendentalist. In the first half of the twentieth century, analytic and positivist philosophers characterized the Indian philosophical spectrum as an assortment of rival claims about causal efficacy and logical accuracy. In this context, Nāgārjuna was viewed as a logical analyst of competing metaphysical and epistemological propositions. Subsequently, postempiricist post-Wittgensteinians have seen Nāgārjuna as an antiphilosopher, primarily concerned with language use, conceptual holism, and the limits of philosophical discourse.

These phases represent far more than ways of understanding the texts of another culture. The Mādhyamika materials under examination, self-conscious in the extreme about the dangers of conceptual presupposition, were not essential to this project, although they were particularly appropriate for an examination of the determining powers of scholarly assumptions and methods. An inquiry of this kind could have been carried out with any major area of scholarly enterprise and any classic text—this study is an inquiry into the philosophy of scholarship.

The hermeneutic insight that scholars are conditioned by social practice and linguistic determinants—not a new insight—does not constitute grounds for dismissing the work of either our predecessors or our competitors as isogetically tainted. But it does offer the suggestion that concepts such as "original context" and "close reading" are simplistically isogetic. A classic is a classic because it engenders multiple meanings. The most lasting truths are found in the least reductive configurations of the largest possible number of conflicting interpretations. In other words, the most useful interpretation may well be one that takes into account as many previous interpretations as possible and attempts to disclose the ways in which these earlier readings made sense, both to the interpretive scholar and to his or her readers.

Rather than contributing one more theoretical discussion of hermeneutics, or offering one more attempt at textual exegesis, this study examines the degree to which specific interpretations of a specific text have been determined by factors often apparent only from the standpoint of another interpretive era or perspective. Furthermore, this study demonstrates the often stated principle that, rather than an ahistorical search for a preferred method or philosophy of interpretation, the enterprise of interpretation is inherently historical. Every reading of a text—including, of course, the most carefully contextualized and historicised readings—will, in some ways, be unavoidably determined by some set of prejudgments. The choice is, therefore, not between good readings, undetermined by irrelevant considerations, and bad readings, rendered inaccurate by interpretive prejudice. The choice between one reading and an even better reading is a difference in degree and not in kind. Within any set of rules for what counts as a desirable interpretation, choices between more and less preferable readings of texts can and will be made. And a study such as this suggests that our conventionally agreed-on rules of interpretation—the rules that tell us what *is* relevant, and what sorts of judgments *are* harmfully prejudiced—are anything but constant. Our preferences in regard to what constitutes a good interpretation are just as determined as our readings themselves.

I thank Victor Preller and Jeffrey Stout at Princeton University for the encouragement as well as the edification I received as their student. I thank my friend, Jeffrey Perl, for his invaluable editing, his scholarly example, and his generous conversation. And I thank my wife, Holly Fairbank, for her support and friendship.

New York A. P. T.
September 1989

CONTENTS

Comparative Philosophy and the Philosophy of Scholarship

Prejudices are not necessarily unjustified and erroneous, so that they inevitably distort the truth. In fact, the historicity of our existence entails that prejudices, in the literal sense of the word, constitute the initial directedness of our whole ability to experience. Prejudices are biases of our openness to the world. They are simply conditions whereby we experience something—whereby what we encounter says something to us. (HANS-GEORG GADAMER, *Philosophical Hermeneutics,* Berkeley, 1976, p. 9.)

Unfortunately it's a question of words, of voices, one must not forget that, one must try and not forget that completely, of a statement to be made, by them, by me, some slight obscurity here. (SAMUEL BECKETT, *The Unnamable,* New York, 1955, p. 384.)

1

The Philosophy of Scholarship

Comparative Origins

In 1786, Sir William Jones, the founder of the Asiatic Society of Bengal, announced that study of the Sanskrit language held the key to the origins of the classical languages of the West:

> The Sanskrit language, whatever be its antiquity, is of a wonderful structure; more perfect than the Greek, more copious than the Latin, and more exquisitely refined than either, yet bearing to both of them a stronger affinity, both in the roots of verbs and in the forms of grammar, than could possibly have been produced by accident; so strong indeed, that no philologer could examine them all three without believing them to have sprung from some common source.[1]

Jones suggested that as well as genealogical connections with Greek and Latin, there were also possible relations with the Germanic, Celtic, and Persian languages, and undeniable similarities between classical Indian and Western mythologies. A "comparative grammatical study" was to be undertaken to determine the full extent of these relations and to enable Jones to complete his official task as Chief Justice at Fort William of codifying Sanskrit legal traditions for the East India Company.

Histories of Indian studies and introductions to books on ancient Indian civilization tend to acknowledge Jones as the "undisputed founder of Orientalism"[2] and as the man "whose efforts in India opened Sanskrit studies to the West."[3] The address of 1786, from which I have quoted, is said to have "ushered in a grand era of comparative philology when Sanskrit and Persian became the keys to unlock the prehistoric world of Indo-European, its parent language whose considerable progeny flourish in many modern languages."[4] And his declaration that Indian and European classical languages were historically and structurally linked is said to have been the initiating impulse for the establishment of Indian philosophy, Indian literature, and comparative philology as legitimate subjects of academic inquiry.[5]

This choice of ancestry on the part of modern scholars is interesting. Jones was not the first European to have studied Indian texts and languages in some depth. In 1651, a Dutch missionary, Abraham Roger, published some of the works of Bhartṛhari along with a book on Brahmanical texts, *Open Door to the Hidden Heathendom.* Fifty years later, the first Sanskrit grammar by a European was written by a Jesuit priest, Johann Ernest Hanxleden, who had served for over thirty years in the Malabar Mission. Subsequently, two more Sanskrit grammars were composed by Fra Paolino de St. Bartholomeo (an Austrian Carmelite whose real name was J. Ph. Wessdin), who also wrote several books on Indian culture that "show a great knowledge of India and Brahmanical literature, as well as a deep study of Indian languages and especially of Indian religious thought."[6] Nor was William Jones the first Englishman to have begun the task of compiling and translating Sanskrit texts. He was preceded in this endeavor by Charles Wilkens, who had studied Sanskrit in Benares for some years before Jones was appointed to his post at Fort William, and who had already begun a translation of the *Institutes of Manu,* which Jones was to assist him in completing. It was Wilkens who gave Europe the first translation of the *Bhagavadgītā* (in 1785) as well as translations of the *Hitopadeśa,* the "Śakuntalā" episode of the *Mahābhārata,* and a Sanskrit grammar for which he carved and cast the type himself to enable publication.[7]

But it is Jones who is seen as the pioneer of Indian studies, and this indicates much about how contemporary scholars regard their own work and define their own disciplines. European interest in India had always been decidedly, almost exclusively, commercial before Jones's statements excited interest in the nature of Hindu language and literature. Prior to the seventeenth century, spices for the tables and kitchens of Europe had been virtually the only object of Western curiosity and desire in South Asia. But after the exclusion of the British from the Dutch-held Indonesian islands, one of the two major sources of spices in the East, they were forced to turn their attention to the other source,

India, and to other marketable Indian goods, primarily printed textiles, silks, and muslin. Encouraged by a growing trade with China and increased European prosperity, Western interests shifted dramatically to India as a source for cheap, brightly colored, and washable fabrics. The second half of the eighteenth century saw the consolidation of British economic hegemony by the East India Company, Jones's employer, and the beginning of British political control over the previously independent Indian states. The entire land mass of India was perceived as economically exploitable British property, and its population was viewed as a source of cheap, or free, labor. This was hardly the climate in which to generate fascination with the treasures of Indian learning and cultural achievement. But it was Jones's insistence on the *comparative* nature of the "Orientalist" enterprise that drew attention to the Sanskrit language and to Hindu culture as objects of intrinsic value, like spices and textiles. This emphasis on comparison is the clue to his acknowledged reputation as a scholarly founding father. He implied that Europe could learn things about itself from India that it could learn nowhere else, and this proved, for scholars, to be a far greater motivation for studying India than learning about an utterly alien—culturally unrelated—civilization. To assert that the study of Indian language could assist European self–understanding was to assert that knowledge about India was of ultimate importance to Europeans, and, consequently, after Jones, Indian subjects could no longer be dismissed as exotica. Within three decades of his address to the Asiatic Society, the first chair in Sanskrit studies at a Western European university was occupied by Antoine-Léonard de Chézy at the Collège Royal de France, Friedrich Schlegel had written his very influential work, *Über die Sprache und Weisheit der Indien,* and his older brother, August Wilhelm Schlegel, had become the first professor of Sanskrit at the University of Bonn. Most pertinent to Jones's philological concerns, in 1918 Franz Bopp published *Über das Conjugationssystem der Sanskritsprache,* a realization of Jones's proposal for a systematic comparison of Sanskrit with German, Greek, and Latin for the purpose of illuminating the origin and the basic structure of all Indo-European languages. All of these scholars explicitly expressed their indebtedness to Jones and agreed that he had succeeded in making Indian language and Indian culture of vital interest to a previously disinterested Europe.

Jones had published an English translation of Kālidāsa's "Śakuntalā" in 1789, the Sanskrit text of another Kālidāsa work, "Ṛtusaṃhara," in

1792, and a translation of the *Institutes of Hindu Law, or the Ordinance of Manu* in 1794. That these texts excited as much widespread European enthusiasm as they did can hardly be explained by their content alone. German poets no less renowned than Herder and Goethe expressed delight at the Sanskrit-to-English-to German translation of "Śakuntalā" (1791), and the German Romantic movement, under the leadership of the two Schlegels, believed that the "appearance of the Sokuntola [sic] . . . was the unfolding of the history of the primeval world which up till now is shrouded in darkness."[8] German translations of Jones's translations and essays circulated widely among nonspecialists and contributed to a fad for Orientalia. The Romantics seized on Indian literature as a direct link with cultural origins and a source of liberation from the stifling atmosphere of rationalism that had dominated European thought in the eighteeneth and nineteenth centuries:

> To the Romanticist, who had become painfully aware of himself in the icy breath of the rationalistic, European-Christian atmosphere of a sobering disengagement from his own roots, India appeared like the promised land . . . there the link of life with the archaic, the contact with the profoundly mysterious and ancient coherence of existence had not been torn apart—the placenta had not yet cut loose.[9]

The nineteenth-century tendency to romanticize Indian literature, and to "discover" answers to European concerns and parallels with European thought, can be seen nowhere more clearly than in the celebrated story of Schopenhauer's encounter with an early and very questionable translation of the Upaniṣads. This translation had begun its colorful history in Delhi in 1656 at the court of the Moghul Prince Mohammed-Dara Shakoh, brother to Aurangzeb and son of Shah Jahan, the builder of the Taj Mahal. Dara Shakoh had brought pandits from Benares to begin the monumental task of translating all Indian religious texts into Western languages. Over a period of two years they did succeed in translating fifty Upaniṣads into Persian.

Almost a century-and-a-half later, in 1801 and 1802, the French philologist, Anquetil Duperron, published a Latin translation of this Persian collection under the title, "Oupnek'hat," a corruption of the word, "Upaniṣad." In 1814, Friedrich Majer introduced Schopenhauer to this "absolutely imperfect Perso-Latin translation of Anquetil Duperron . . . full of misinterpretations and not the Upaniṣads as we know and explain them now."[10] From that moment on, Schopenhauer expressed his indebtedness to Majer for opening up the world of Hindu thought to him and he enthusiastically asserted his own close philosophical kinship with the ancient authors of the Upaniṣads. He proclaimed this Latin translation of a Persian version of an incomplete set of Sanskrit texts to be "the production of the highest human wisdom," and in the preface to *Die Welt als Wille und Vorstellung* he acknowledged not only that his own work was inspired by this "Oupnek'hat," but that "if the reader has already received and assimilated the divine inspiration of ancient Indian wisdom, then he is best of all prepared to hear what I have to say to him."[11] Elsewhere in his writings he stated that whoever reads the Persian-to-Latin translation will be "gripped to his innermost being . . . every line is full of firm, positive, and consistently coherent significance."[12] He claimed that it was "the most rewarding and inspiring text in the world," that "it has been the consolation of his life and will be still at his death," and that its authors, "can hardly be conceived of as having been mere men."[13] He is famous for his boast that, "did it not sound too conceited, I might assert that each of the individual and disconnected utterances that make up the Upaniṣads could be deduced as a consequence from the thought I am to impart, although conversely, my thought is by no means to be found in the Upaniṣads."[14] Schopenhauer saw no problem in asserting that the authors of ancient Indian texts were attempting to answer precisely the same questions that had troubled Immanuel Kant and himself. He identified the Upaniṣadic term, *Brahman,* with the Kantian *Ding an Sich* and read the Upaniṣadic maxim, *Tat tvam asi* ("That art thou"), as a direct expression of the Indian belief in the transcendental unity of all reality and a corresponding belief in the phenomenal world as a manifestation of the constituting activity of the transcendental ego.

Schopenhauer's appropriation of the Upaniṣads for his own purposes was by no means an exception to common practice, though it is probably the most notorious. This practice was widespread and unquestioned. Throughout the nineteenth century European scholars consistently grafted their own intellectual concerns and discursive practices onto an India that was virtually of their own creation and treated Indian texts as exotic expressions of their own presuppositions and philosophies. In his study of European attitudes toward Islamic culture, *Orientalism,* Edward Said has argued to the contrary and claimed that Europe has consistently pictured Asia as "one of its deepest and most recurring images of the Other . . . the Orient has helped to define Europe (or the

West) as its contrasting image, idea, personality, experience."[15] And although it is true in many cases that Europeans have portrayed Asia as a dark, threatening, ultimately unknowable anti-Europe, it is equally the case that the urge to find parallels, to see Asia as a mirror, has been at work, particularly among those scholars who are professionally engaged in the translation and interpretation of ancient Indian texts.

These scholars claim descent for their disciplines from the comparative philologists who saw India as a source of information about European origins. Their interests in India derive not from fascination with the ways in which it is culturally alien to Europe, but from the belief that the two cultures are linguistically and philosophically consanguineous. In his "Discourse on the Philosophy of the Asiatics," William Jones argued that Europeans could not even claim exclusive possession of religious teachings that were customarily considered to be of Christian origin. He cited passages from Confucius and from the *Hitopadeśa* that were, in his estimation, the Asian versions of "two Christian maxims . . . to do to others as we would they should do unto us, and to return good for evil,"[16] and that certainly predated Christ by several centuries. He asserted that, along with the linguistic parallels linking Europe and Asia, there were equally compelling philosophical and religious relations. To a great extent, the history of modern Indological studies is a history of that comparative compulsion.

Isogesis and the Ideal of Objectivity

That modern South Asianists still identify their work with Jones's comparative enterprise is particularly striking when one considers another of their central ideological commitments—the criterion of objectivity. Interpretations and translations of ancient texts are intended to be as "accurate," "objective," and "close to the original" as possible. This assumption has existed, of course, not only in Sanskrit scholarship, but in virtually all fields involving historical, textual, and cross-cultural understanding, and has roots in the traditional (nineteenth-century) hermeneutic theory associated with Schleiermacher and Dilthey. The standard reading of the nineteenth-century hermeneuticist position is that it is concerned with the recovery of original textual meaning, which can be recovered only through the reconstruction of the historical, psychological, and cultural context in which the text had been written. In this view, it is held that the interpreter's personal beliefs block true understanding of the original author's words and must be transcended if the interpreter is accurately to "recover the original life-world they betoken and to understand the other person (the author or historical agent) as he understood himself."[17] In essence, traditional hermeneutics insists that there is one true meaning that is the goal of any interpretation, and that this meaning is effectively identical with the author's intention. The interpreter's job is to set aside his own cultural, historical, and personal biases and to retrieve this objective meaning by entering the world of the author.

There is a tension here. William Jones's assumption, that Asian materials are crucially interesting because of what they can tell us about ourselves, clashes with the methodological goal of exegetical objectivity. If the acknowledged purpose of the textual investigation is to shed light on one's own culture and language, it is difficult to imagine that the investigator can go about selecting and explicating textual materials without cultural bias. The comparative enterprise is inherently subjective. But no translator or scholar engaged in textual exegesis wants to think that he is *guilty* of reading his own cultural presuppositions, or forcing his own interests onto the text under investigation:

> To impose our own categories on the data provided by the Buddhist source materials is to run the risk of violating their intentionality and, consequently, to vitiate the entire interpretive enterprise.[18]

The very idea of "reading into" a text is anathema. But at the same time these interpreters, trained in a discipline that has its roots in cross-cultural comparison, value knowledge about the foreign as a way to better understand their own culture. This eighteenth-century, Jonesian enthusiasm for comparison has ever since been a formal bias of Indian studies, and it mixes only up to a point with the nineteenth-century hermeneutical passion for objectivity.

The dissonance between subjectivity and objectivity in textual interpretation is the subject of intense debate in other fields in which subjectivity is assumed, notably literary criticism and contemporary philosophical hermeneutics. However, even in Indian textual studies, in which the emphasis is on philology and translation, the practical effects of this methodological tension are always inescapable. The criterion of objectivity has engendered an exegetical style. But it is undeniable that readers of Indian texts unwittingly engage in a kind of *isogesis,* a "reading into" the text that often reveals as much about the interpreter as it does about the text being interpreted. Isogesis is an unconscious phenomenon, whereas exegesis is simply conscious intent. I will not want to argue that an obviously isogetical reading, like that of Schopenhauer, is necessarily evil or necessarily good, nor that an old-fashioned hermeneuticist, following the objectivist prescriptions of Schliermacher or Dilthey, would necessarily do better or worse. Instead I argue throughout this study that isogesis is inevitable in all readings of texts: it is not a failure of understanding but the evidence of it.

The nineteenth-century zeal for objectivity in interpretation has not blunted the eighteenth-century appreciation for self-knowledge that is also a part of the Indianists' pedigree. Nor has it provided a methodology for textual interpretation that avoids the imposition of European practices of discourse onto readings of another culture's literature. Western interpretations of Indian texts, particularly the philosophical texts, have been and will continue to be strongly comparative. European scholars have consistently looked in the Indian intellectual tradition for answers to Western philosophical problems. They have used European technical terminology in translations and analyses of Sanskrit texts, and in the interests of elucidating Asian thought for a Western readership, they have made it an accepted practice to compare Western with Indian philosophers.

In themselves, parallels between European and Indian thinkers are not conceptually troubling and I am not implying that the comparative emphasis in Indian studies automatically invalidates any scholarly works. There is little reason why comparisons cannot or should not be made. Difficulties do arise when, for strictly comparative purposes, a text, a writer, or a philosophical school is lifted from its appropriate historical, cultural, and intellectual context. But it is clear that the more contextual the treatment, the "thicker" the description,[19] the less chance there will be of overt misrepresentation. Less obvious, however, is the significance of tone. It is a very different thing to assert that both Kant and the author (or authors) of the *Bṛhadāraṇyaka Upaniṣad,* for example, are "saying the same thing" than to demonstrate the possibility of reading the *Bṛhadāraṇyaka* in a Kantian manner, and the divergence between these is a distinction between different comparative tones. This difference in tone can also be described as a distinction of degrees of comparative enthusiasm. It is the difference between declaring that something essential and immutable has been "discovered" about the ideas contained in these texts, on the one hand, and, on the other, constructing a context in which two intellectual traditions can be understood together. This seemingly trivial distinction between the tones of discovery and patient, responsible construction is, in fact, crucial, and hinges on the degree of self-consciousness of the scholarly interpreter. There is surely nothing philosophically incorrect or offensive about wanting to claim that two particular philosophers from different cultural traditions both seem to be grappling with essentially the same issue and that they appear to be doing basically the same thing with it. Many cross-cultural theoretical comparisons are of this inoffensive sort and are the evidence of a creative attempt on the part of a scholar to read one philosopher coherently into another philosophical context. But the temptation to declare that both the Kantian and Upaniṣadic views contain the same truth, although it is perhaps often irresistible, is inherently problematic if one is also concerned with issues of the indeterminacy of meaning, with the difficulties associated with traditional theories of truth as correspondence, and with the impossibility of reading from a culturally or historically neutral standpoint. To argue for a transcendent sort of philosophical identity is to implicate oneself in arguments for determinate meaning and philosophical universality and to ignore arguments that the interpreter is also necessarily included in the hermeneutic process. When one says (with the tone of discovery, rather than construction) that two philosophers from separate traditions are "saying the same thing," "have the same position," or are attempting to "solve the same problem," one edges close to the belief that philosophical problems are universal, perennial puzzles with which all men, regardless of historical period or culture, are engaged, and that these eternal problems can be expressed in any language and at any time. On the other hand, many contemporary, postempiricist thinkers argue that philosophical problems are closely and necessarily tied to mutable, culturally shaped factors, and that the changing of vocabularies and assumptions often brings about the formulation of new philosophical concerns. It would seem impossible, from this position, to even want to claim that Kant and the authors of the Upaniṣads agree on an issue or are "saying the same thing" except in the most casual way. It would be enough to offer a coherent reading of an Indian text in Western terms, or vice versa. And indeed, the kind of comparison that makes minimal claims for itself allows greater intelligibility of interpretation: it gives us all we can have, and that is no small thing.

It is well that scholars in a field such as Indian studies recognize the limits of cross-cultural comparison. But concern with the constraints placed on the interpretive process by historical and cultural factors can become excessive. Faced with a choice of belief in subjectivity or belief in objectivity, scholars in the past century have often tended to choose the latter, and have become in the process victims of a less obvious but no less isogetical unself-consciousness. In the interests of scholarly objectivity and philological accuracy, hermeneuticists argued that scholars must carefully work from "within" the culture or historical period under investigation and divorce themselves from the theories and sensibilities of their own age and culture. South Asianists who were influenced by these arguments have attempted to construct commentaries, translations, and histories that were impervious both to changing trends and to longstanding cultural assumptions in the West:

> Why is it so important to be able to see Hinduism through Hindu eyes, to see the tradition from the inside? For one thing, doing so is a prophylactic against superimposing alien and inappropriate conceptualizations. But more profoundly, this empathetic identification is prerequisite to seeing meaning in the objects of study.[20]

Carefully eschewing European philosophical terminology and limited comparisons with Western thought, European and Indian scholars in the first half of the twentieth century wrote multivolumed "Histories of Indian Philosophy" and "Outlines of Indian Philosophy" that were believed to be authoritative because they were intended to be "objective."[21]

But this notion of "objectivity" is itself a product of Western theoretical assumptions. It is a reflection of both the traditional hermeneuticist and positivist commitment to "the ideal of the autonomous subject who successfully extricates himself from the immediate entanglements of history and the prejudices that come with that entanglement."[22] This commitment, neo-Kantian in nature, brought about the assumption, among scholars in the "human sciences," that an interpreter's own prejudices and beliefs must be neutralized if he is accurately and objectively to understand the thought of another culture. But the Kantian ideal of the neutral observer, like the Cartesian program of doubting all accepted beliefs, assumes the possibility of an epistemologically neutral state, a way of seeing the world that is not influenced by any specific cultural or personal factors.

This ideal is inherently problematic. It ignores the fact that "no man wholly escapes from the kind, or wholly supasses the degree, of culture which he acquired from his early environment,"[23] and it runs counter to the fact that knowledge can be expressed and understood only in specific, culturally embedded forms. There are no non-culture-specific languages in which to write, or unconditioned perspectives from which to view, another age or culture. Consequently, Indianists of the objectivist type face the same difficulties as objectivists in other fields who try to construct "theory-neutral" vocabularies and describe complex phenomena in terms that are meant to be unconditioned by cultural, psychological, and linguistic factors. While trying to counter the subjectivist excesses of comparative scholarship, textual objectivists become guilty of a different kind of methodological excess.

Put another way, scholars for the past two centuries have defied the isogetical nature of their work by attempting to put aside their own prejudices and presuppositions. By attending as carefully as possible to lexical questions, historical detail, and the accumulation of more and yet more texts to translate and interpret, they have created, for themselves as well as for their readers, an illusion of a progressive increase in knowledge about Indian philosophy and of a steady accumulation of better readings of Indian philosophical texts. The standard belief has been that there are more or less correct interpretations of texts and that the meaning of a text is recoverable if all of the necessary philological and historical research is carried out. Concomitant with this belief is the view that disputes between interpreters can be adjudicated, and that there are ways of finding "correct" readings that are not dependent on the assumptions of the interpreter. Deficiencies in textual interpretation are understood to be a result of "an imperfect acquaintance with primary source materials"[24] and it is assumed that greater familiarity with original texts and the restriction of the scholar's modern Western biases will give us "accuracy" and greater understanding of Indian thought. This type of textualist positivism has been reinforced by the view that the interpretation of another culture's texts is primarily a philological matter and that the production of a good translation is tantamount to solving most important interpretive questions. What questions remain can be answered by paying close attention to historical and cultural context and avoiding the "superimposition" of our own habitual biases.

It is not surprising that this type of objectivism was extremely influential throughout the late nineteenth and early twentieth centuries, when positivisms of all sorts were common in philosophy and the social sciences, but it is extraordinary that readers of Indian texts still continue to be naively concerned with discovering the "real meaning" of texts. As Jeffrey Stout stated:

> Opposition to meanings has become so fashionable in recent years that some writers speak of the "heyday of meanings" as a bygone era, a thing of the past.[25]

In many scholarly fields, terms such as "objectivity," "meaning," "truth," and "intention" have been seriously questioned: they can no longer be used without inverted commas. The writings of Heidegger, Wittgenstein, Quine, Davidson, Sellars, and Kuhn have been celebrated for their attacks on various types of philosophical essentialisms. The Anglo-American New Criticism of William Wimsatt, Cleanth Brooks, F. R. Leavis, and William Empson has severely curtailed interest in authorial intention among writers on Western literature. And the more recent textualists and "postcritics" as diverse as Barthes, Said, Foucault, Derrida, Gadamer, Bloom, and Rorty are so influenced by all of the aforenamed that they perceive the notions of literal meaning and truth as representation only as remnants of antique theoretical premises:

> they are saying . . . that a certain framework of interconnected ideas—truth as correspondence, language as picture, literature as imitation—ought to be abandoned. They are not, however, claiming to have discovered the *real* nature of truth or language or literature. Rather, they say that the very notion of discovering the *nature* of such things is part of the intellectual framework we must abandon—part of what Heidegger calls "the metaphysics of presence," or "the onto-theological tradition."[26]

In literary theory, the end of the nineteenth century saw the beginnings of a widespread dissatisfaction with the notion of the "meaning" of a literary text, just as the values and assumptions of "realism" were being questioned by the new "modernist" arts. And in philosophy, by the 1950s, critiques of "explanation," "analyticity," and "meanings" had begun to seriously erode the status of "truth" in interpretation. Dilthey's hermeneutical principle that the meaning of a text is to be discovered by uncovering the original intentions of the author has been disputed by any number of theorists of interpretation. Hans-Georg Gadamer's major work, *Truth and Method,* is one such attack:

> Every age has to understand a transmitted text in its own way, for the text is part of the whole of the tradition in which the age takes an

> objective interest and in which it seeks to understand itself. The real meaning of a text, as it speaks to the interpreter, does not depend [solely] on the contingencies of the author and whom he originally wrote for. It certainly is not identical with them, for it is always partly determined also by the historical situation of the interpreter and hence by the totality of the objective course of history . . . Not occasionally only, but always, the meaning of a text goes beyond its author. That is why understanding is not merely a reproductive, but always a productive attitude as well.[27]

It is the "productive" or creative aspect of interpretation that I refer to as isogetical. There are no interpretations that are not the result of some creative effort on the part of the interpreter, and it is difficult to imagine what would be gained from an interpretation that did not exhibit the isogetical interference of the commentator:

> An interpretation involving no creativity of this kind would be uninteresting and purposeless, for it could consist in nothing more than repetition of the text itself. Readings are either creative or superfluous.[28]

That scholars in an interpretive field like Indian philosophical studies have attempted to deny their own creative input and to claim that their efforts are totally "objective" reveals a curious combination of excessive modesty, hubris, and ignorance (or less insultingly, innocence). To believe that an interpreter is nothing more than a mouthpiece for the original author, that he is acting only as an "interface" between the world of the text and the world of the reader, is to humbly downplay the creative component in the act of interpretation. But simultaneously, for an interpreter to believe that he *can* accurately reconstruct the intentions and beliefs of the original author without betraying his presence is nothing less than belief in his own scholarly omnipotence. Most surprisingly, for contemporary Indologists to naively accept nineteenth-century objectivist principles betrays an ignorance of the methodological debates that have been taking place throughout the twentieth century in the closely related fields of literary criticism and postpositivistic European/American philosophy.

There are present-day supporters of textual objectivism who have tried to defend Dilthey's and Schleiermacher's interests in the intentions of the author; E. D. Hirsch's *Validity in Interpretation* is a prominent example of this kind of defense. But many Indologists still write as if there have never been questions raised about the validity of reconstructing an author's private intentions. They sometimes acknowledge the difficulties that a European may have in attempting to understand "the Indian way of thinking." They believe that much historical and cultural research is necessary to adequately capture the intentions of an ancient non-Western author—but the basic goal of objective interpretation is rarely questioned. Their concern is with explicating the meanings intended by the Buddha in the sixth century BC, or Śaṃkara in the eighth century AD, or Rāmānuja in the eleventh century AD. There is normally little or no attention paid to the fact that, as interpreters, they will necessarily work their own creativity into the texts, and that the notion of textual understanding cannot be reduced to the simple epistemological relation between the subject, an interpreter, and the object, a text.

If we are to reject the idea that meaning lies hidden beneath the surface of a text, then we have to see the act of reading as an enterprise that involves the context of the reader as much as that of the text. Readers of Sanskrit texts, like readers of European novels, must employ their personal and cultural perspectives if they are to find what they read intelligible. This act of productive understanding—isogesis—is an integral part of the interpretive process, and the putative objectivism of many scholars in Indian studies is only a vain attempt to deny the phenomenon.

Philosophical Fashions in Indian Studies

The natural sciences undergo dramatic shifts in theory and professional practice—"revolutions"—that bring about entirely new ways of defining the tasks of science and the ways in which scientists see the world. This at least has been the argument of Thomas Kuhn and his followers. The thesis has met with strong resistance in some areas because it appears to challenge deeply held beliefs about scientific progress and the reliability of scientific knowledge. It surprises no one that such paradigm shifts are common occurrences in "softer" fields that are believed to employ the "less rigorous" investigative techniques of textual interpretation and philosophical speculation. The enterprise of textual interpretation is subject to trends, fashions, and even fads, and it should not be controversial to assert that scholars are always contemporary. They are strongly influenced by prevailing assumptions, values, goals, problems, terminologies, and methodologies. In the case of Indian philosophical studies, this means that, as the assumptions of modern interpreters have changed, their readings of the Indian tradition have changed accordingly. A text that is central to the interests of one generation of scholars may be drastically reinterpreted or virtually ignored by a subsequent generation.

The last century has brought significant variation, decade by decade and commentator by commentator, in the interpretation of Indian texts—and there is little to suggest that this variation is wholly a result of the increase in our knowledge of the Indian tradition. Although much historical and philological information has in fact accumulated, shifts in interpretation and evaluation are, more often than not, a result of radical shifts in the philosophical orientation of the interpreters. Thus, much of what Kuhn has said about scientific change in the West is also applicable to shifts in the Western readings of Indian philosophical texts. Richard Rorty summarizes this view in his discussion of the general nature of philosophical change:

> Interesting philosophical change (we might say "philosophical progress" but this would be question-begging) occurs not when a new way is found to deal with an old problem but when a new set of problems emerges and the old ones begin to fade away. The temptation (both in Descartes's time and in ours) is to think that the new problematic is the old one rightly seen.[29]

In each generation, the new problematics of Western philosophy have yielded correspondingly new, but not necessarily more "correct" readings of the Indian tradition. And each new set of readings has reflected the assumptions of each new generation of readers.

It was not until the late nineteenth century that "Indian philosophy" was first recognized as an independent subject for scholarly inquiry. Prior to this time, the treatment of the Indian philosophical systems, as a field of study distinct from Hindu religion and literature, was virtually nonexistent: most early nineteenth-century studies of Indian culture treated philosophical, poetic, and religious literatures as equal and indistinguishable objects of philological research. Indian philosophical ideas were of only passing interest to the scholars who worked with these texts in the original languages, since they were trained as linguists and not as philosophers. But the lack of specialized interest in Indian philosophical matters on the part of European scholars may have resulted as much from their own perception of "philosophy" as from their limited knowledge of, or interest in, Indian philosophical texts. In the nineteenth century, philosophy came to be perceived in Europe as an autonomous academic discipline, distinct both from theology and from the natural sciences:

> It was not until after Kant that our modern philosophy-science distinction took hold. Until the power of the churches over science and scholarship was broken, the energies of the men we now think of as "philosophers" were directed toward demarcating their activities from religion. It was only after that battle had been won that the question of separation from the sciences could arise.[30]

Once the Kantian shift from metaphysics to epistemology had been accomplished, philosophy could be viewed as a "field" of inquiry distinct from all others, a professionalized area of expertise with its own concerns and techniques. For Sanskritists, the result of the professionalization of philosophy in European universities was a rush to demarcate the "strictly philosophical" in their own field. Differentiations began to be made among Indian texts: some were labeled products of poetic or religious inspiration, and others were offered as examples of pure philosophy. With the publication of works such as Paul Deussens's *The System of the Vedānta* (1883), Richard Garbe's *The Philosophy of Ancient India* (1897), and Max Müller's *Six Systems of Indian Philosophy* (1899), Indian philosophical study was established as an independent subdiscipline within Sanskrit studies, and the subdiscipline had a distinctly Kantian flavor.

In other words, the founders of "Indian philosophical studies" conceived their enterprise under the influence of eighteenth- and nineteenth-century German idealism. (Among his numerous other philological enterprises, Müller had translated Kant's first critique into English.) Their writings on Indian texts are infused with Kantian and Hegelian terminology, neo-Kantian beliefs about the primacy of epistemology, and idealist concerns with "transcendental Truth." Consider this selection from Paul Deussens's study of the Upaniṣads:

> The whole of religion and philosophy has its root in the thought that (to adopt the language of Kant) the universe is only appearance and not reality *(Ding an Sich);* that is to say, the entire external universe, with its infinite ramifications in space and time . . . only tells us how things are constituted for us, and for our intellectual capacities, not how they are in themselves and apart from intelligences such as

> ours . . . There have been three occasions, as far as we know, on which philosophy has advanced to a clearer comprehension of its recurring task, and of the solution demanded: first in India in the Upanishads, again in Greece in the philosophy of Parmenides and Plato, and finally, at a more recent time, in the philosophy of Kant and Schopenhauer.[31]

Deussen enthusiastically joined other idealists in their redefinition of philosophy as a study of the disjunction between appearance and reality. His readings of Upaniṣadic and Vedāntic texts were meant to disclose ancient Indian philosophy as a tradition devoted to uncovering the noumenal reality obscured by the transcendentally constituted veil of phenomena. Deussen asserted that "all great religious teachers . . . are alike unconsciously followers of Kant"[32] in that they agree on the proposition that all empirical reality, all workings of causality in the realm of space and time, is mere appearance. Phenomenal reality is not

> a disposition of "things in themselves," to use Kant's words; [it] is mere *māyā* and not the *ātman*, the "self" of things, as the Upaniṣads teach. For in this case there is room for another, a higher order of things set over-against the reality of experience, from the knowledge of which we are excluded by our intellectual constitution, which religion comprehends in faith by her teaching concerning God, immortality, and freedom. All religions therefore unconsciously depend on the fundamental dogma of Kantian philosophy, which in a less definite form was already laid down in the Upaniṣads. These last therefore by virtue of their fundamental character lie naturally at the basis of every religious conception of existence.[33]

German idealism presented a lens through which the Indian philosophical tradition appeared to have been duplicating the latest "discoveries" of the great European thinkers. And British idealism, as represented by the works of F. H. Bradley and T. H. Green, held similar sway over Anglo-American Indianists. In the West, interest in Indian thought was engaged as idealists of all kinds began to see compelling parallels between Śaṃkara's *nirguṇa Brahman* and such Western concepts as Kant's *noumena,* Fichte's *Ich,* Hegel's *Geist,* Bradley's Absolute, and later, Heidegger's *Dasein.* Buddhist scholars of an idealist bent found their own parallels: Dignāga's distinction between the levels of pure sensation and cognitive inference was compared with Kant's distinction between *noumena* and *phenomena,* the Yogācāra *ālaya-*

vijñāna was identified with the Hegelian Absolute Idea. It seemed that all of India saw reality as a "motionless whole."

During this initial, idealist phase, Indian thought was portrayed as an ongoing competition between six "orthodox" schools or views *(darśanas):*

1. The Sāṃkhya (said to be founded by Kapila)
2. The Yoga (Pantañjali)
3. The Mīmāṃsā (Jaimini)
4. The Vedānta or Uttara Mīmāṃsā (Bādarāyana or Vyāsa)
5. The Nyāya (Gautama)
6. The Vaiśeṣika (Kaṇāda).

From the very outset, this "six schools" convention, acquired directly from contemporary Indian authorities, was both universally accepted by European scholars and taken by these same scholars to be inherently problematic. This traditional classificatory scheme was based on the Brahmanic position that the Vedas were the foundation of all Indian thought. Any system that did not at least nominally acknowledge the scriptural authority of the Vedas was to be regarded as heterodox and outside the scope of pure Indian philosophy. As a result, the contributions of major non-Hindu movements—such as Buddhism and Jainism—to the history of Indian philosophical dialogue were disparaged or overlooked. Similarly other, less well known but historically documented, schools of thought, such as the Cārvākas (materialists), Ājīvikas (fatalists), and Lokāyatikas (naturalists) could be discussed only briefly as insignificant heresies or philosophical aberrations. It was the received view among European scholars that there were six major systems of Indian philosophy—no more, no less—and this view allowed strange inconsistencies and omissions to occur in their studies.

For example, in his work of 1875, *Indian Wisdom,* Sir Monier Williams (best known today as the compiler of the standard Sanskrit–English dictionary) included three chapters meant to provide an overview of Indian philosophical thought. After naming the six famous schools and their supposed founders, he went on to devote more than half of his discussion to discounting the heresies of the Buddhists, Jains, and Cārvākas, and to suggesting that the six philosophical systems were not really six in number. Instead, they were

> practically reducible to three, the Nyāya, the Sāṃkhya, and the Vedānta. They all hold certain tenets in common with each other and to a

> certain extent also (especially the Sāṃkhya) with heretical Buddhism.[34]

In other words, the standard "six systems" view, promulgated by Brahmin pandits to enhance their own prestige and reduce the notoriety of their historical opponents, was passively accepted by European scholars and passed on to their readers even though they found it difficult to support through their own research. The Yoga system could be described, not as a philosophical system, but only as the psychological–therapeutic application of the Sāṃkhya school. The Vaiśeṣika system was said to be virtually identical with the Nyāya. And the Mīmāṣā school was usually dismissed as no more than the ritualistic codes that went along with Vedānta teachings. Even in a book titled *The Six Systems of Indian Philosophy,* Max Müller was unable to present a history of Indian thought that consisted of six discrete and autonomous sets of teachings, and that did not refer continually to the heterodox beliefs of Buddhists, Jains, and others. Clearly, the Brahmin classificatory system—which divided philosophies into those that did not accept the authority of the Vedas *(nāstika)* and those that did *(āstika)*—was acting primarily as a theoretical encumbrance.[35] Although it accurately reflected the teachings and prejudices of modern Hindu scholars, it did little to help Western scholars create a consistent picture of the history of ancient Indian thought.

But the six-schools theory provided something more important than consistency for nineteenth-century idealist scholars. It provided a schematic grid that they could lay over the Indian sources. Most scholars believed, with Monier Williams, that there were really only three significant Indian philosophical positions: the Nyāya, the Vedānta, and the Sāṃkhya. The decision to drop certain schools from primary consideration (the Yoga and the Mīmāṃsā) and to coalesce others into a single school (the Nyāya-Vaiśeṣika) was intuitive. There was little question that the Mīmāṃsā emphasis on religious ritual made its teachings "less" than philosophical, that the Yoga view was something other than rational argumentation, and that the Nyāya and Vaiśeṣika schools agreed on the "important issues." They accepted a philosophical school as significant only if it took a position on what they believed was the central issue of philosophy: the disjunction between appearance and reality. They accepted that there were six schools but their interest was entirely in three.

From the start, the Sāṃkhya school, usually acknowledged as the oldest of the *darśanas,* was labeled as "radical dualism." It provided a "Cartesian" foundation for the Europeanized history of Indian philosophy. The *Sāṃkhya-Kārikā* spoke of two distinct ultimate realities—*prakṛti,* the physical world or matter, and *puruṣa,* the self or spirit—a division that European idealists readily identified with the familiar Western distinction between object and subject. All aspects of the physical world, including the physical elements of a human being, were said to be a manifestation of *prakṛti,* which was devoid of all consciousness. Correspondingly, *puruṣa* was the "self" or "soul" that existed alongside the physical, but that was not itself a material entity. *Puruṣa* was to be understood as pure consciousness, a passive witness to the changing, active, material world of *prakṛti.* Furthermore, there was not one *puruṣa,* but many: the Sāṃkhya contended that there must be as many *puruṣas* as there were distinct and unique conscious individuals. They rejected any suggestion that *puruṣa* could be reduced to *prakṛti,* or vice versa. Experience, as the Europeans agreed, is always constituted of two distinct poles, the experience itself and the experienced, the awareness and the object of awareness. Consequently, the distinction between *puruṣa* and *prakṛti,* between the mind and the body, was, for the Sāṃkhya, absolute and indissoluble.

The goal of Sāṃkhya philosophy was man's liberation from the imperfections and limitations that arise out of *puruṣa's* involvement with *prakṛti.* To Western readers it seemed that the Sāṃkhya position was self-contradictory. On the one hand, there was the claim that the imperfections of the self were due to its entanglement with *prakṛti;* on the other, there was the equally important claim that the evolution of *prakṛti,* the process by which the material world takes shape and changes, is the only means by which the self can liberate itself from those same imperfections. The self, due to ignorance, falls victim to the illusion that its involvement with *prakṛti* is real: that salvation consists in the realization that such involvement and the attendant imperfections are unreal and result from a false identification with *prakṛti.* Thus, liberation, in the Sāṃkhya scheme, is a process by which the *puruṣa* comes to comprehend its eternal and absolute independence from *prakṛti.* It is a process that can take place only through manipulation of the world of *prakṛti* by the *puruṣa*—through the practice of Yoga.

It was not difficult for the Sāṃkhya viewpoint to be interpreted by Western scholars as a facsimile of familiar Platonic and Cartesian concerns. It appeared to offer versions of the traditional Western philosophical distinctions between appearance/reality, mind/body, and subject/object. But the Sāṃkhya solutions to the perennial problems of philosophy were perceived by these same Western scholars as unsatisfactory: the Sāṃkhya school was characterized as a primitive and unsuccessful attempt to solve dilemmas that could be properly treated only by a more mature, less "dualistic" philosophy. As they viewed it, Sāṃkhya's failure was due to its portrayal of the world as an unstable equilibrium between mutually exclusive realities. *Prakṛti* and *puruṣa*—two totally independent and yet integrally related patterns of the real—presented a tension that was internally inconsistent. For nineteenth-century readers, accustomed to an ongoing conflict between empiricists and idealists, materialists and rationalists, positivists and transcendentalists, there could be no single school of thought that successfully incorporated both views. In their estimation, consistency in philosophical thought demanded a simple resolution to the conflict. The choice was between a material, objective reality or a subjective, transcendental one. The Sāṃkhya school wanted both and consequently looked to Western scholars like a doomed, archaic hybrid. Kantian readers demanded a distinction between *noumena* and phenomena. *Prakṛti* and *puruṣa* were both noumenal, both ultimately real.

In the historically more recent Nyāya-Vaiśeṣika (about 400 BC), Western idealist interpreters thought they had found the "objectivist" side of the Sāṃkhya dualism. The Nyāya was, in their estimation, the Indian version of European empiricist realism, complete with meticulous analyses of inference and an emphasis on perception as the most valid source of knowledge. The Naiyāyikas were portrayed as the consummate logicians of India and as believers in the most extreme kind of metaphysical realism. All of experience was classifiable into discrete categories. Both particulars and universals were considered ultimately real. Even the human "soul" *(ātman)* was objectified as a discrete particular. Nyāya-Vaiśeṣika methodology was analytical, its metaphysics particularist *(viśeṣas* literally means "particulars"), and its empistemology realistic. According to the Nyāya, the world consisted entirely of individuals, which contain both uniqueness and universal properties that stand to the particulars in a relation of inherence. The Nyāya realists claimed that universals were as real as the particulars in which they inhered. Most significantly, they insisted on the reality of all things that could be classified as objects and rejected the ultimate validity of the subject, particularly in any role as a constituting agent of the phenomenal world. In this way, the Nyāya-Vaiśeṣika school was understood as the ancient Indian opposition to transcendental idealism.

As a consequence, although much admired for its attention to logic,[36] the Nyāya school inspired little philosophical sympathy or creative effort among a modern idealist readership. In 1919, Arthur Keith wrote that

> While the philosophy of the Vedānta is well known in Europe, the Nyāya and Vaiśeṣika, the Indian systems of logic and realism, have attracted hardly a tithe of the interest due to them as able and earnest efforts to solve the problems of knowledge and being on the basis of reasoned argument . . . even historians of Indian philosophy like Professors F. Max Müller and P. Deussen have contented themselves with sketches which ignore entirely the serious and valuable thought of the schools. The result is gravely embarrassing for any serious study of Indian philosophy as a whole. . . . [37]

As Keith points out, the focus of European attention during the late nineteenth and early twentieth century was overwhelmingly on Vedānta. Although there were virtually no full-length studies of Nyāya-Vaiśeṣika or Sāṃkhya written by Europeans,[38] there were numerous titles on Vedānta, particularly on Śaṃkara's Advaita Vedānta. Keith believed that the neglect of the realist schools could be explained by

> the nature of the original sources . . . modes of expression unequalled for obscurity and difficulty . . . their details frequently defy explanation, and in translation are more obscure if possible than their originals.[39]

But the Nyāya-Vaiśeṣika texts were not more obscure or inaccessible than those of the Vedānta, they were merely less interesting to European idealists. Müller devoted most of his own research on Indian philosophy to the Vedas, Upaniṣads, and Vedānta, and stated that the Nyāya and Vaiśeṣika systems

> are very dry and unimaginative, and much more like what we mean by scholastic systems of philosophy, businesslike expositions of what can be known.[40]

In contrast, he wrote that, out of all the Indian schools, it was only the Vedānta philosophers who were engaged in a "search after truth."[41] Other philosophical systems were "very inferior in interest to the Vedānta," which had "a decided priority in importance."[42] In the preface to his *Six Systems of Indian Philosophy* he went so far as to say that the

Vedānta philosophy is "a system in which human speculation seems to me to have reached its very acme."[43]

In this climate of ardor and receptivity, Indian–Europe comparisons were inevitable. One of Paul Deussen's articles on Vedānta (1893) was titled "The Philosophy of the Vedānta in Its Relations to the Occidental Metaphysics,"[44] and was published in the *Journal of the Asiatic Society of Bengal,* the official organ of William Jones's foundation. We have already seen evidence of Deussen's comparative intentions in his discussion of the Upaniṣads—his works on Vedānta (most prominently, *Das System des Vedānta,* 1883) were similarly devoted to demonstrating the presence of Kantian ideology in Hindu texts. Müller's *Three Lectures on the Vedānta Philosophy* (1894) followed the same normative program. To European idealists, "non-dualism," or *advaita,* was the obvious solution to the material/spiritual dualism of the Sāṃkhya. The *Brahmasūtras* stated that "those who have realized the *Brahman* have realized it as their own *ātman.*"[45] And in his commentary to those sutras, Śaṃkara asserted that "*Brahman* is a state of being wherein all distinctions between self, world, and God are transcended and are obliterated."[46] The Vedāntic doctrine of *Brahman/ātman* identity convinced European scholars that Vedānta was identical with German transcendental idealism, especially when combined with consistent denial of the reality or reliability of perceptual experience (the concept of *Māyā).* In opposition to Nyāya-Vaiśeṣika's realism and empiricism, Vedānta philosophers refused to admit the reality of the world of appearance and posited only a transcendental world of spirit. Any apparent multiplicity in perceptual experience was dismissed as illusory, and all diversity or change was explained as a transformation of the energies of the single, universal *Brahman.* In Kantian terms, *Brahman* was the "thing-in-itself," the ground of phenomenal appearance. If the Nyāya school represented scientific empiricism to European readers, the monistic Vedānta could be nothing other than an Indian version of the idealism of Kant, Schopenhauer, Fichte, and Hegel. Deussen's enthusiasm was boundless:

> On the tree of Indian wisdom there is no fairer flower than the Upaniṣads, and no finer fruit than the Vedānta philosophy. This system grew out of the teachings of the Upaniṣads, and was brought to its consummate form by the great Śaṅkara (born 788 A.D., exactly one thousand years before his spiritual kinsman Schopenhauer). Even to this day, Śaṅkara's system represents the common belief of nearly all

thoughtful Hindus, and deserves to be widely studied in the Occident.[47]

Vedānta came to represent all Indian thought for Europeans during the next several decades. Sāṃkhya and Nyāya attracted little or no scholarly attention, but the notion of a subcontinent of idealists captured the Western imagination.

As more works on Indian philosophical subjects were produced during the following years, their character gradually changed in two significant ways. First, there was a change in the interpretive tools and interests of Western Sanskritists, and second, there was a concomitant shift in the object of their attentions. The first change, to be more specific, may be characterized as an alteration of philosophical orientation on the part of the European scholarly community. In the first few decades of the twentieth century, idealist biases began to give way to newer philosophical trends, as new approaches, following Frege and Russell, took hold. Some years later, terms such as "sense data," "meaning," "reference," and "denotation," along with a new and intense interest in linguistic and logical analysis, began to appear in the Indianist literature. Although attention to Indian idealisms continued, the new analytically oriented scholars, in the manner of the Vienna Circle, shifted their focus away from the primarily metaphysical interests of the idealists and began to concentrate more on Indian views about language, logic, causation, and the justification of knowledge. Unlike the idealists, the analysts did not overtly favor any one Indian *darśana* to the exclusion of all others. In fact, their concentration on technical philosophical issues "for their own sake" brought about a serious reevaluation of the merits of the "six-schools" classification system and encouraged a general broadening of scholarly attention to a much larger spectrum of Indian philosophical positions.[48]

This general widening of scope did not preclude, however, a concentration of interest on specific "analytic" schools and texts. The greatest interest among Westerners centered on the realists, pluralist, and logic-centered systems such as Nyāya-Vaiśeṣika and the early Buddhist schools, which analysts found to exhibit an almost Russellian attention to the logical form of philosophical propositions. Moreover, these realist and particularist schools of Indian thought, formerly neglected, seemed to twentieth-century scholars to correspond more closely to modern scientific realism and logical atomism than did the now out-of-date idealist monisms. Analytic and positivist philosophers could associate their own efforts with those of Indian philosophers, who appeared to see particular linguistic entities, linked by ostension, to particular objects in the world, and who tried to learn about the world through "meaning analysis."

Defenders of the traditional idealist bias in Indian studies eyed this "analytic turn" with suspicion and saw it as a resurgence of the materialistic objectivism that Kant and Hegel had all but completely eradicated:

> I am not unaware of the fact that in fashionable contemporary philosophical circles a reaction has set in of late against absolute idealism, of which Bradley was a prominent representative . . . Absolutism, it will be said, is an outmoded doctrine. Well, so far as contemporary British philosophical thinking is concerned . . . its dominant note is scientific empiricism . . . [But] the predominant trend of philosophical thinking in India has always been the absolutist Vedānta, either of the school of Śaṃkara or that of Rāmānuja and it continues to be so even to this day, though in some circles of philosophical thinking in contemporary India which have remained solely under the influence of Western philosophy, the spell of British empiricism has begun to work.[49]

Pride in the idealistic philosophies of the Upaniṣads and Advaita Vedānta was very important to Asianists, as is obvious from the tone of the lines above. It had become an established belief that all Indian thought was devoted to the attainment of *mokṣa,* or liberation from the "earthly bondage" of perceived reality. This "otherworldliness," the idealism that had so impressed Sanskritists such as Müller and Deussen, had been a cherished possession. The new "analytic" approaches to Indian thought, and the accompanying attention to Indian realisms, threatened the "old-fashioned conception that India was and is the land of dreamers and mystics."[50]

A new generation of Sanskritists and philosophers began to argue that absolutism (or idealism) had not been the only significant voice to be heard in the history of Indian thought. Scholars of that history, Western and Indian, began to argue that materialism, naturalism, and empiricism had been as highly developed in Asia as in Europe. In *Indian Philosophy, A Popular Introduction,* Debiprasad Chattopadhyaya (the author also of *Lokāyata, A Study in Ancient Indian Materialism)* contended that

> The emphasis so far on the idealist trend, to say the least, has been lopsided. The acceptance of the Advaita Vendānta, for example, is even looked upon as a mark of philosophical respectability . . . But in our traditional philosophy itself there were also vigorous attempts to outgrow idealism.[51]

New histories and outlines of Indian philosophy were written without the usual emphasis on Vedānta and idealism.[52] Full-length studies of heterodox, antiidealist schools were produced for the first time.[53] And a number of new studies organized around specific philosophical topics—such as causation, theory of meaning, and logical strategy—reinforced the tendency to concentrate on Indian realism.[54] Modern Western philosophers had shifted their allegiance from subjective idealism to "critical realism," "logical positivism," "logical empiricism," and "conceptual analysis." Within a very few years, it began to seem as if classical Indian philosophers had also been concerned with theories of logic, meaning, and perception, with justification of belief, philosophy of science, and linguistics.

In the second half of this century, the analytic turn in interpretation was joined by an alternative mode of philosophical discourse. Beginning with Wittgenstein's assault on the viability of traditional "theory of knowledge" and on the efficacy of meaning analysis, and continuing with Quine's critique of the distinction between language and fact, with Sellars' attack on the "Myth of the Given," with Davidson's holistic reevaluation of "truth," and with Kuhn's and Feyerabend's portrayals of scientific facts as "theory-laden"—the notion that language can correctly represent or "picture" the world by connecting words with objects, sense-data, or facts has lost currency with a new kind of philosopher. Holistic antifoundationalism has also begun to have an effect on the Indianists' interpretive work. This trend, which I shall refer to, somewhat reductively, as "post-Wittgensteinian," has, like the idealist one, favored a particular philosophical system (Mādhyamika Buddhism) and has proceeded by inventing parallels (between Nāgārjuna and Wittgenstein).

This new comparison centers on the claim that both Wittgenstein and Nāgārjuna offer a functionalist view of language ("meaning as use") and that their reservations about any linguistic expression of "Truth" lead to a parallel skepticism about the utility of every philosophical formulation. It has become possible to describe Wittgenstein's prescription for a philosophical "therapy" as a search for something not unlike Buddhist enlightenment, and to portray Nāgārjuna as an Indian metaphilosopher, concerned with "language-games" and "forms of life." Wittgenstein, it has been asserted, "was applying to European absolutism the same critique earlier applied to Indian absolutism by the proponents of the Middle Way";[55] moreover, "all modern adherents of the Mādhyamika ought to be Wittgensteinians."[56] The claim has been made that "much of what the later Wittgenstein had to say was anticipated about 1800 years ago in India,"[57] and, as if that were not enough, that "only a Wittgensteinian interpretation will suffice for certain central Buddhist concepts. . . . Buddhist philosophy once took a markedly Wittgensteinian turn."[58] Even the most sober of comparisons states that "Nāgārjuna's use of words for articulating Ultimate Truth would find champions in contemporary philosophers of the language analysis school such as Ludwig Wittgenstein or P. F. Strawson."[59] In recent years, there has been a marked increase in the publication of books and articles on Mādhyamika: there is now a growing number of Wittgenstein-influenced Sanskritists and Tibetanists writing studies on the works of Nāgārjuna, Candrakīrti, Bhāvaviveka, and Tsong Khapa. Contemporary studies of Vedānta (of which there are far fewer than there were at the turn of the century) concentrate on explaining the differences between the philosophies of Śaṃkara and Nāgārjuna.[60] (Fifty years ago, Nāgārjuna would not have been mentioned at all.)

The post-Wittgensteinian climate obviously encourages appreciation of the Mādhyamika texts, and it encourages the same degree of isogetic enthusiasm that was evident in Müller's taste for Vedānta and in Karl Potter's for the Nyāya. Though the study of Indian philosophy has gone through several distinct interpretive fashions since its inception over a century ago, the basic tendency to project Western philosophical concerns into Sanskrit texts has never been discarded. In fact, this habit of "reading into" the text (which can be traced most immediately to William Jones and the early Asian philologists, but which is undoubtedly a part of all textual interpretation) is noticed only when an old interpretive trend begins to give way to a new one—such is always the case with changes in fashion. It would be inaccurate to suggest that the roughly characterized phases of idealist, analytic, and post-Wittgensteinian interpretation have followed each other in a neat historical sequence, or that any one approach has been immediately abandoned after the appearance of its "successor." There are, even in the Wittgensteinian heyday, idealist and analytic scholars still at work. New schools and new trends are more novel than new: the basic pattern, which is isogesis, has not changed in over 100 years. The new interpretive readings of the twentieth century—Sarvāstivādins who sound like Russell, Naiyāyikas who agree with Carnap, or Prāsaṅgika-Mādhyamikas who appear to be Sellarsian—are the products of minds that see different philosophical problems, use different philosophical vocabularies, and write in vastly different styles than do their Kantian, Hegelian, or Bradleyan colleagues. Nineteenth-century idealists had believed that truth was to be found in nineteenth-century ideas. Twentieth-century positivists discovered their kind of "hard facts" and "rigorous logic" in their own analyses of Indian texts. And postphilosophical Wittgensteinians have found a second-century Indian Buddhist who questions the validity of twentieth-century philosophical analysis.

The history of Indian philosophical studies is a history of isogetic interpretations. This is clear, but is it significant? This episode in the history of scholarship draws its significance from the incongruous pairing of comparatist and objectivist imperatives: those modern demands of all the interpretive disciplines. Together they are profoundly incompatible and, consequently, expectations have not been realized. Scholarship has looked to theory for solutions to its problems, while, of those problems, theory has been, more often than not, the source.

2

Nineteenth-Century German Idealism and Its Effect on Second-Century Indian Buddhism

The Rehabilitation of Mādhyamika

Almost a thousand years ago, Indian Buddhism had been reduced to ashes—untold numbers of Buddhist monks slaughtered, monasteries razed, texts destroyed—during the Afghan and Turkish invasions. As a result, European travelers and merchants found no Buddhists, and few remains of Buddhist culture, when they first established regular contact with India at the end of the fifteenth century. For the West, Buddhism was an exclusively East Asian phenomenon. Even three centuries later, the great European Orientalists were convinced that the study of Buddhism had little significance for their investigations of Indian history and religion. It was thought that Buddhism might have originated in India, as is evident in this passage from a seventeenth-century study of Asian history:

> The origine of this religion, [Buddhism] which quickly spread thro' most Asiatick countries to the very extremities of the East . . . must be look'd for among the Brahmines.[1]

But eighteenth- and nineteenth-century scholars were confident that Buddhism had not been an important factor in Indian cultural history, nor indeed a legitimate religion. William Jones himself had dismissed the Buddha as a "bastard manifestation of the Egyptian sun god,"[2] and Henry T. Colebrooke, the man who, after Jones, did the most to establish the character of English Sanskrit studies for the remainder of the nineteenth century, followed suit. Colebrooke referred only cursorily in his writings to the "heretical systems of Jina and Buddha" and he relied entirely on Sāṃkhya and Mīmāṃsāka sources for his negative characterizations of Buddhism as "atheism" and as a philosophy of "perfect apathy."[3] Even Brian Hodgson, who is known for presenting the first major collections of Buddhist texts to Western Europeans, wrote that

> I had no purpose, nor have I, to meddle with the interminable sheer absurdities of the Buddha philosophy or religion.[4]

Because of a chronic respiratory problem, Hodgson had been unable to remain in his diplomatic post in India, where he had first been assigned by the British government. Instead he took up residence in the higher altitudes of Nepal for a period of ten years, where he had hoped to continue his research into ancient Hindu texts. During this period, 1883 to 1843, Hodgson came on a large number of unfamiliar Sanskrit manuscripts and began to send copies to libraries throughout Europe. These texts, which are now known to belong to the Mahāyāna tradition, were all but ignored by European specialists in Indian studies, who were interested chiefly in Hindu sources. But one French scholar, Eugène Burnouf, of the Société Asiatique of Paris, began to see the outlines of a complex system of thought in the manuscripts being sent to Europe by Hodgson.

In the first half of the nineteenth century, Burnouf was acknowledged as Europe's premier master of Sanskrit and Pāli, and he was the first to investigate seriously the Buddhist texts written in those languages. There had been European Buddhologists who had preceded Burnouf, such as Jean Pierre Abel Rémusat (1788–1832) and Isaac Jacob Schmidt (1799–1847), but they were almost exclusively concerned with the still living traditions of East Asian Buddhism and the Chinese, Mongolian, and Tibetan translations associated with them. Burnouf's studies, however, were the first to center on Buddhism as an Indian phenomenon and the first to develop a connection between the early texts of the Pāli canon and the later Sanskrit texts of the Mahāyāna schools. His concerns were initially with chronology and with establishing the primacy of Indian languages in the history of Buddhist literature. But along with an attention to historical and philological matters, Burnouf offered Europe an interpretation of significant Buddhist terms that has exerted influence to the present day. He translated *nirvāṇa* as "extinction," and asserted that the goal of Buddhist philosophy was

> a disappearance of individuality by way of absorption—into the Supreme Being or into the void (*śūnyatā*) . . . in any event, *nirvāṇa*

> means a fundamental change in the condition of the individual, that would, to all appearances, be utter annihilation.[5]

Burnouf's *Introduction à l'histoire du Buddhisme indien* (1844) and his annotated translation of the *Saddharmapuṇḍarīkam* (*Le Lotus de la bonne loi,* 1852) firmly established the idea throughout Europe that Buddhism was a religion of negativity and nihilism. Burnouf's documents belonged principally to the *Prajñāpāramita* (or "Perfection of Wisdom") tradition and included Candrakīrti's *Prasannapadā,* a major commentary on Nāgārjuna's *Mādhyamikakārikā*. These texts impressed Burnouf chiefly as expressions of *Śūnyavāda,* or "the doctrine of the void," and he saw, in Nāgārjuna's thought in particular, the clear expression of an unqualified philosophical nihilism. Max Müller, who had been a student of Burnouf's, wrote that

> no person who reads with attention the metaphysical speculations of *Nirvāṇa* contained in the Buddhist canon can arrive at any conviction different from that expressed by Burnouf, *viz.* that *Nirvāṇa,* the highest aim, the *summum bonum* of Buddhism, is the absolute nothing.[6]

Other writers, notably Jules Barthelemy Saint-Hilaire, the journalist and translator who was to be proclaimed by Müller as "the first true historian of Buddhism," read Burnouf's speculations about the nature of *nirvāṇa* and went even further toward developing the negative "annihilationist" reading of Buddhism. Saint-Hilaire viewed Buddhism as "a monstrous enterprise in which every potential service to mankind is sterilized by a pervasive nihilism,"[7] and made this prediction:

> I believe that the study of Buddhism . . . will show how a religion which has at the day more adherents than any other on the surface of the globe, has contributed so little to the happiness of mankind; and we shall find in the strange and deplorable doctrines which it professes, the explanation of its powerlessness for good.[8]

The prevailing nineteenth-century view of Buddhism was that it was a philosophy of negativism, and the Orientalists responsible for introducing Buddhism to Europe (Burnouf, Müller, Saint-Hilaire) were the first to dismiss it. Saint-Hilaire, in particular, decried the Buddhist tradition as "horrible and naive,"[9] sarcastically remarking that "the Buddha was merely the most logical and audacious of the Hindu philosophers."[10] The only justification for studying Buddhism at all, he added, was that it helped one to "appreciate the abiding values of our own heritage."[11] His vituperative attacks, joined with Burnouf's more cautious philologically oriented misgivings, formed a consensus that was not to be challenged for many years, though tremendous amounts of new textual material, in both Pāli and Sanskrit, were being added to European collections of Buddhist manuscripts.

The few defenders of Buddhism—men like Albrecht Weber, Jean Baptiste Francois Obry, and Philippe Foucaux—argued that Buddhism had operated as a force for social reform, fostering notions of equality and change in socially repressive Brahmanic India. They avoided discussion of philosophical issues and asserted that doctrinal differences between Buddhist schools demanded that each tradition be evaluated separately: that Buddhism should not be condemned as a whole. They tried to defend against the charge of nihilism either by ignoring the entire issue or by sacrificing one Buddhist school, inevitably Nāgārjuna's Mādhyamika, to the opponents, hoping to exonerate the rest in the process. It was not until the second decade of the twentieth century that anyone thought to argue that Mādhyamika could be taken seriously, that it might be understood as anything besides the most extreme of irrationalist nihilisms. For all of these scholars, the question was whether the entire spectrum of Buddhist schools, or Mādhyamika alone, should be branded as "annihilationist." In other words, as Guy Welbon indicates in his survey of European Buddhology,

> Burnouf's opinion about Nāgārjuna and the prajñāpāramita literature was the standard until the great dispute between Louis de La Vallée Poussin and Th. Stcherbatsky. That is to say, no nineteenth-century author attempted to make a case for the Mādhyamikas as non-nihilists.[12]

To explain Saint-Hilaire's antipathy, Welbon notes that

> In Barthelemy Saint-Hilaire's discussion of Buddhist *nirvāṇa,* one sees a great deal more personal involvement displayed than in the commentaries written by earlier European students of Buddhism. . . . All Barthelemy Saint-Hilaire's judgments of Buddhism were buttressed by his deep and informed commitment to the philosophical, emotional, and religious foundations of his Western European society. Classical Greek and Christian traditions, he believed, had sanctified and fructified human life. From his point of view, all Indian thought denied the values fostered and supported by Christianity and the philosophies of ancient Greece.[13]

But the European antipathy toward Buddhism as a whole, and toward Mādhyamika in particular, can be explained only by something other than simple religious intolerance. As already observed, although Buddhism was being vilified throughout Europe, Hindu philosophy was being celebrated by the same scholars as the height of intellectual achievement. There was no shortage of appreciation for things Indian: Buddhism was singularly derided.

The Buddhologists' terminology of "annihilation," "extinction," and "nihilism" discloses an obsession with the metaphysical and a mistrust of any system that bypasses questions of ontology.[14] Burnouf's principal concern was whether the individual is ultimately "absorbed" into a transcendental ground of existence or totally "extinguished" in *nirvāṇa.* And for all the scholars who followed him, the suspicion that the Buddhist *nirvāṇa* was in fact nothing—that there was no realm of reality behind the veil of appearance—made European acceptance of Buddhist philosophy impossible. Burnouf, Saint-Hilaire, and Müller saw nothing but annihilation of the individual, a denial of the reality of what Kantians were calling the "transcendental ego," what Hegelians were calling *Geist.*

Transcendental idealism, although it did shift the focus of philosophical discourse from metaphysics to epistemology, did little to eliminate the longstanding Platonic assumption that an ontological foundation was necessary—that there must actually be something casting the shadows onto the wall of Plato's cave. In the first *Critique,* Kant did argue that human beings could know nothing of the thing-in-itself, but it was also essential that things-in-themselves could be said to exist in some form. However much an idealist asserted that all of experience was constituted by the forms and categories of the understanding, there was always the existence of the constituting agent, the transcendental ego, that was foundational. Without at least this existential ground, there could be no epistemology. Without some ontology, philosophy was nonsense. Buddhism, however, particularly Mahāyāna Buddhism, appeared to suggest that existence itself was an illusion, that the ego, transcendental or otherwise, was no more real than any other concept or intuition. Belief in *nirvāṇa* demanded belief that there was no "real *noumena* grounding all "unreal" *phenomena.* So with a consensus that Buddhist philosophy had little relevance to the conclusions of the great European philosophers, Buddhist studies remained in the hands of philological specialists who continued to collect and translate texts from

Pāli and Sanskrit. Mādhamika, commonly agreed to be the most nihilistic of all Buddhist schools, was ignored.

In the first decades of this century, however, the Belgian Buddhologist, Louis de La Vallée Poussin, found himself, unexpectedly, in heated dispute with a Russian scholar of Sanskrit and Tibetan texts. Fyodor Ippolitovich Stcherbatsky, founder of the "Leningrad school" of Buddhist scholarship, advanced a controversial position about the nature of *nirvāṇa* and the Mādhyamika texts. La Vallée Poussin, like Burnouf before him, had labeled Mādhyamikas as "pure nihilists,"[15] but, on other grounds, saw himself as a great champion of Mahāyāna thought. He had argued, against such well-known scholars as T. W. Rhys Davids and Hermann Oldenberg, that the Pāli sources were far from sufficient for understanding Buddhist philosophy and he had stressed the necessity of studying the later Sanskrit and Tibetan texts. He was responsible for producing a definitive edition of Nāgārjuna's *Mūlamadhyamaka-kārikās,*[16] with Candrakīrti's *Prasannapadā,* from three separate Sanskrit versions and one Tibetan translation. La Vallée Poussin had suggested, in his controversial study, *Nirvāṇa* (1925), that genuine Buddhism, which in his estimation did not include Mādhyamika, was not inherently nihilistic. The Buddhist goal, he argued, was not annihilation, but a condition of eternal bliss, a passage of the immortal soul into a transcendent paradise. Mādhyamikas, who seemed to deny the reality of both the soul and eternity, were for him extremists who advocated a heretical departure from Buddhist orthodoxy.

For his part, Stcherbatsky saw La Vallée Poussin's suggestion that the Buddha had preached a doctrine of belief in an existing soul as nothing more than a European apologetic for Buddhism. It showed, in his words, "much more what [La Vallée Poussin] desired Buddhism to be than what it really was."[17] Against La Vallée Poussin, Stcherbatsky defended the standard scholarly opinion that *nirvāṇa* was annihilation, but argued a position that no European scholar of Buddhism had yet thought to suggest. He was the first to write that regardless of soteriological emphasis and negative terminology, no Buddhist school, not excluding Mādhyamika, was guilty of philosophical nihilism. Stcherbatsky's primacy is part of the standard history:

> From the time of Burnouf through the most famous writings of Louis de La Vallée Poussin, Europeans have been generally agreed that, whatever *nirvāṇa* might signify, there could be no doubt that the *śūnyatā* (so-called voidness) theory of the Mādhyamika stamped that

> school as completely nihilistic. Stcherbatsky was almost alone in insisting that the Mādhyamika's denial of the reality of the dharma moments was not nihilism. *Śūnyatā* meant "relativity," he declared, the voidness of the interdependent moments in themselves.[18]

Stcherbatsky's earliest publication, *Theory of Knowledge and Logic in the Doctrine of the Later Buddhism* (1903), was expanded in 1930 to become his magnum opus, *Buddhist Logic* (1930), and it offered an idealist interpretation of late Yogācāra philosophers that departed significantly from all earlier Western readings of Buddhist thinkers. Moreover, his later writings, the critical review of La Vallée Poussin's *Nirvāṇa* (1925) and *The Conception of Buddhist Nirvāṇa* (1927), argued vociferously that Nāgārjuna's philosophy of *śūnyatā* was not an expression of nihilistic excess. According to Stcherbatsky, Nāgārjuna's emphasis on "emptiness" was meant to deny the apparent reality of empirical phenomena but not to deny the reality of the thing-in-itself—the absolute. He asserted that Mādhyamika was a philosophy of the ultimate oneness of all reality: it was, in his terms, a "radical monism." To argue for this controversial interpretation, Stcherbatsky produced the first major Kantian-influenced readings of Buddhist texts.[19] In the process, he brought to European Buddhist studies the first serious philosophical treatment of Nāgārjuna's thought.

Appearance and Reality

Kant had made a careful distinction between "concepts" and "intuitions" in his first *Critique*. At the beginning of the "Transcendental Aesthetic," he asserted that there are two necessary conditions for any kind of experience. First, there must be something directly given or "presented" to the senses—intuitions. Second, these intuitions must be organized, by what he termed "concepts of the understanding," into various categories. Only after the intuition has been synthesized by one or more concepts, recognized as of a certain kind or as having certain properties, can there be said to be conscious experience. Without concepts, pure intuitions would be useless:

> Thoughts without content are empty, intuitions without concepts are blind.[20]

In Stcherbatsky's studies of the writings of Dignāga (fourth century) and Dharmakīrti (seventh century),[21] this same Kantian distinction is offered as the foundation of a Buddhist epistemology. According to these Indian philosophers, Stcherbatsky asserted, the "senses" alone apprehend the world directly, while the "intellect" constructs a synthetic world of conceptual illusion. The distinction between intellect and sense, like Kant's dichotomy of concept and intuition, forms the basis for a theory of how the mind creates a world of continuity, consistency, order, and substance. For the Buddhist logicians, Stcherbatsky explained, there are two levels of reality, the conditioned or empirical one (*saṃvṛti-satya*), produced by the cognitive intellect, and the ultimate or absolute reality (*paramārtha-satya*), which is immediately perceived in something called "pure sensation." In this way, Stcherbatsky's explication of the Buddhist theory echoes, but subtly alters, Kant's disjunction between *phenomena* and *noumena*. Stcherbatsky's Buddhist, like Kant, does not admit the ultimate reality of the objects of normal cognition, and refers to them only as mental constructs. But unlike Kant, Dignāga and Dharmakīrti contend that it is possible to attain an awareness of the absolute: in other words, it is possible for the human mind to encounter noumenal existence. These Mahāyāna Buddhists appeared to offer a Kantian epistemology without the limitations that Kant placed on metaphysics. Like Schopenhauer, Müller, and Deussen, Stcherbatsky had discovered an Indian version of Kantian philosophy, but one that differed markedly from the Vedānta, the Hindu school that had engaged the attention of other Indianists. For the first time, a European scholar read Buddhist texts in a way that seemed philosophically acceptable in the post-Kantian climate.

In *Buddhist Logic,* Stcherbatsky's explication of Dignāga's and Dharmakīrti's thinking is organized around two related principles: the epistemological theory of "pure sensation" and the metaphysical theory of "instantaneous being." The theory of sensation explains the method whereby ultimate reality can be cognized, whereas the theory of instantaneous being describes the nature of ultimate reality itself. Real cognition is to be found only in the act of "new" cognition, the initial perception of the object not yet fully cognized. After that point, any sensation of an "object" is, in reality, only "recognition." Stcherbatsky wrote that for this school of Buddhist philosophy, there are no enduring objects and no enduring sensations of any object:

> In every cognition there is a sensible core and an image constructed by the intellect, one part is sensible, the other intelligible. The thing itself

> is cognized by the senses, its relations and characteristics are constructed by imagination which is a function of the intellect.[22]

After the initial instant of cognition, the continued perception of an object is actually an act of what Stcherbatsky termed "judgment." In any cognition there are these two distinct phases: the first moment of perception and the subsequent phase of inferential judgment:

> What is cognized by the senses is never subject to cognition by inference, and what is cognized by inference can never be subject to cognition by the senses.[23]

For the people that Stcherbatsky refers to as "realists" (by which he means the Buddhists' opponents, principally the Nyāya-Vaiśeṣika philosophers), the visual perception of a fire (the favorite Indian rhetorical example) is a clear instance of simple sense perception. When the same fire is unseen but its existence is assumed because of the perception of smoke, the realist says that one apprehends reality through inference. For the Buddhist, however, perception and inference are involved in both instances. In the first case, in which fire is actually seen, there is indeed pure sensation, but the realization that there is fire exists only on the level of inference. In the second case, the instantaneous cognition of smoke is pure sensation; the immediate conceptualization of smoke and subsequent impression of fire are both instances of inference.

With this doctrine of two, and only two, sources of knowledge, there is one source (sensation) that directly intuits ultimate reality, and another (inference) that constructs the images in which reality is arranged in the phenomenal world. In the case of perception, objects are cognized directly; in the case of inference, they are cognized indirectly, that is abstractly. In viewing the phenomenal world, one performs an act of synthetic judgment, fusing the pure sensation of the thing-in-itself with the thought-constructed object of inference. One sees an essence, and thinks "cow"; one does not see a "cow." Stcherbatsky wrote that the Sanskrit term that he translated as "judgment" (*adhyavasāya*) is, in its normal usage, a "decision."[24] So if a judgment (decision) contains the immediate sensation of the objects and its subsequent conceptualization, then there is, in every perception, a direct cognition of the absolute as well as an abstract decision regarding the categorization of the perception. As would be obvious to any Kantian, these abstract concepts of the intellect are relatively unreal as compared to the "thing-in-itself." It is only the transcendental, preconceptual objects of sensation that can be considered noumenal. Like all good realists, the Naiyāyika philosophers accepted the reality of the subject, and of the object and its qualities. In contrast, Stcherbatsky's Buddhist asserts a dual system of reality and attributes ultimate reality only to the transcendent, unintelligible nonconcept.

The Buddhist's ultimate reality can be perceived and described, and it is at this juncture that Stcherbatsky's brand of Buddhist Kantianism departed from standard German idealist strategies. Dignāga's and Dharmakīrti's metaphysical principles, described by Stcherbatsky as "The Theory of Instantaneous Being" (*Kṣaṇika-vada*), with its correlate theory of "The Falsity of Extention," describe the world as a continuous, fluctuating assemblage of "point-instants":

> The theory of Universal Momentariness implies that every duration in time consists of point-instants following one another, every extension in space consists of point-instants arising in contiguity and simultaneity, every motion consists of these point-instants arising in contiguity and in succession. There is therefore no Time, no Space, and no Motion over and above the point-instants of which these imagined entities are constructed by our imagination.[25]

The ultimate reality of point-instants is, Stcherbatsky insisted, a reality of pure causal efficiency. No duration, extension, or movement is admitted by the Buddhists, simply what Stcherbatsky termed "momentariness." From one isolated instant to the next, there is nothing that, in itself, remains continuous. What we perceive as movement is actually a series of "frozen instants," connected through causality, but without temporal or spatial continuity ("To exist means to exist separately"):[26]

> Ultimate reality is non-constructed, non-imagined, non-related reality, the thing as it strictly is in itself, it is the mathematical point-instant.[27]

To connect his theories of sensation and momentariness, Stcherbatsky claimed that the metaphysical reality of the point-instant is the reality that is intuited through direct sensation. The reality of the Buddhist logicians is a discontinuous flow of causal entities, the mathematical point-instants, which exist only to the extent that they "do something." They exist in a web of efficiency, disappearing as soon as they appear, but "causing" the next moment to occur by their momentary presence. This cinematic picture of ultimate reality is attributed, with little alteration, to other Buddhist schools, and is also the centerpiece of

Stcherbatsky's 1923 study of early Buddhist (Abhidharma) philosophy, *The Central Conception of Buddhism and the Meaning of the Term "Dharma"*:

> The elements of existence are momentary appearances, momentary flashings into the phenomenal world out of an unknown source. . . . They disappear as soon as they appear, in order to be followed in the next moment by another momentary existence. Thus a moment becomes a synonym of an element (*dharma*), two moments are two different elements. An element becomes something like a point in time-space . . . the elements do not change, but disappear, the world becomes a cinema. Disappearance is the very essence of existence; what does not disappear does not exist. A cause for the Buddhist was not a real cause but a preceding moment, which likewise arose out of nothing in order to disappear into nothing.[28]

Stcherbatsky's two major studies devoted to the early and late Buddhist schools from which I have just been quoting, *The Central Conception of Buddhism* and *Buddhist Logic,* present related expositions of metaphysical pluralisms. Early Buddhism, represented by the philosophy of the Abhidharma, was, according to Stcherbatsky, based on the idea of an interdependent multiplicity of discrete elements or *dharmas* ("The conception of a *dharma* is the central point of the Buddhist doctrine").[29] This first period of Buddhist thought was labelled by Stcherbatsky as a "radical Pluralism," having as its main doctrines the denial of the reality of the soul (*anātma-vāda*), the denial of the reality of substance (*dharma-vāda*), and the ruling principle of causation (*pratītya-samutpāda*) that produces the illusion of a stable, material world. The early Buddhistic philosophy, like the later school of Buddhist logicians, was said to describe reality as a flux ordered by causal laws of relative dependence. There was, according to both of these views, no single, real, material, or spiritual entity that acted as a causal substratum for the formation of the universe.

Stcherbatsky's version of Buddhist history was borrowed from the Tibetan historian Buston (1290–1364), and nominally adopted Buston's "three phases of Buddhism" framework: a pluralistic Abhidharma, a monistic Mādhyamika, and an idealistic Yogācāra. But his actual portrayals of the first and third phases are virtually identical. He presents both early Abhidharma and late Yogācāra philosophy as versions of Kantian idealism, laced with Bergsonian "creative evolution," Newtonian physics, and French *pointillism*. The first volume of *Buddhist*

Logic alone contains more than 100 explicit references to Kant, as well as numerous others to Hegel, Bradley, and Bergson. Terms such as "thing-in-itself," "synthetic *a priori*," and "Absolute" appear on virtually every page of the text and there is no indication that Stcherbatsky felt the need to justify his use of idealist vocabulary. From the analytic high ground of a later era, Richard Robinson criticized Stcherbatsky's methodology for precisely this reason:

> Studies have translated a non-Western system into a Western one and have misrepresented this destructured, untextured changeling as a description or explanation of the original. Th. Stcherbatsky's Kantian and Hegelian paraphrases of Indian Buddhism are the most brilliant and flagrant instance of this type.[30]

Stcherbatsky's fondness for his Kantian *pointillism* is evidenced by the tendency in his work to discount the earlier Yogācāra schools (normally associated with Asanga and Vasubandhu) and to carefully explicate only the later Buddhist logicians that fall most easily into line with his own version of Buddhist philosophy. In the introduction to *Buddhist Logic,* he summarized the standard Yogācāra view as a belief in the ultimate reality of consciousness. These philosophers, he said, posited certain classes of ideas as both relatively and absolutely real, and asserted that all existence was necessarily mental. The "store-house consciousness" (*ālaya-vijñāna*) was held to be the ultimate substratum for this mentalistic reality and directly observable through the process of introspection.[31] Although he clearly felt that they were historically necessary for the eventual development of the positions of Dignāga and Dharmakīrti, Stcherbatsky paid the early Yogācāra philosophies minimal attention and few compliments. He characterized Asanga and Vasubandhu as "extremists,"[32] and took pains to side with the philosophers that he termed "logicians" against both these "consciousness-only idealists" and the Nyāya-Vaiśeṣika "naive realists." Like other Europeans before him, he lavished his scholarly attentions on those systems that seemed to echo his own philosophical beliefs—there is little doubt that he found the Buddhist pluralisms to his liking.

I have gone on at length about Stcherbatsky's presentation of Buddhist philosophers other than Nāgārjuna because, although his book on Mādhyamika might appear, on the surface, to be very different from these other major works, Stcherbatsky's treatment of Mādhyamika is formally identical to his interpretations of other Buddhist schools. It is, moreover, most interesting when viewed as a continuation of those earlier projects on the Indian pluralisms. For Stcherbatsky, Nāgārjuna presented a special challenge. His thought could not be readily interpreted, like Dignāga's or the Abhidharma's, as a philosophy of causally interconnected "evanescent elements"—Nāgārjuna specifically denies the reality of independent entities.[33] Nor does it immediately appear that Nāgārjuna allows for an easy epistemological division between concepts and intuitions—he asserts that there is no difference whatsoever between absolute and conventional truth.[34] It would have been very simple for Stcherbatsky, like his predecessors, to dismiss Nāgārjuna as a perverse nihilist, as the philosopher of "nothingness." But he did nothing of the kind.

In *The Conception of Buddhist Nirvāṇa,* first published in 1927 (4 years after *The Central Conception of Buddhism,* 24 years after the initial version of *Buddhist Logic*), Stcherbatsky attempted to explicate "the second phase of Buddhism" in a way that would allow sympathetic European readings of the Mādhyamika school. This project was motivated in large part by his ongoing, energetic debate with La Vallée Poussin about the meaning of the term *nirvāṇa.* Stcherbatsky wanted to show that even the most skeptical of all Buddhist positions could escape the charge of nihilism without being turned into a simplistic faith in heavenly salvation, as he had accused La Vallée Poussin of doing. He went about this by applying European idealist categories and concerns to Mādhyamika texts (specifically, small portions of Nāgārjuna's *Mādhyamikakārikā* and Candrakīrti's *Prasannapadā* commentary) to show that there was, in spite of Nāgārjuna's negativistic terminology of *śūnyatā* (emptiness), a coherent Mādhyamika position on the relation between *saṃsāra* and *nirvāṇa.* In his earlier writings, Stcherbatsky had already (following Buston) referred to Mādhyamika as a "radical monism," and he made no attempt to alter that characterization. In fact, he used the label of monism to turn Nāgārjuna's apparent nihilism into an absolutist idealism. In agreement with La Vallée Poussin and the Polish Buddhologist Stanislaw Schayer,[35] Stcherbatsky asserted that Nāgārjuna's "emptiness" was the "something" that was ultimately real in the Mādhyamika system. But he argued against La Vallée Poussin's interpretation that *śūnyatā* might be simply another name for a Buddhist heaven of eternal bliss and claimed instead that *śūnyatā* must be seen as something like Śaṃkara's *nirguṇa Brahman*—the "real ground of existence" on which the world of conventional reality is (erroneously) founded. By interpreting Nāgārjuna's thought as a transcendental monism he contended that it was possible to avoid "the mistake of regarding his philosophy as nihilism or pure negativism."[36]

Stcherbatsky went so far as to say that "there is but little difference between [Mādhyamika] Buddhism and Vedānta . . . a circumstance which Śaṃkara carefully conceals."[37] Both Nāgārjuna and Śaṃkara, argued Stcherbatsky, denied the efficacy of empirical cognition for realizing the Absolute, and both saw the universe as one "motionless whole where nothing originates and nothing disappears."[38] Mādhyamika, he stated, repudiated causality and "mercilessly condemned" all logic. Soteriological considerations favored mysticism and revelation, since this system saw the world as composed of two realities—one illusory and the other truly real but unknowable through rational or empirical means. This Mādhyamika absolute, or *nirvāṇa,* could be cognized only through "the mystic intuition of the Saint and the revelation of the new Buddhist Scriptures, in which the monistic view of the universe is the unique subject."[39] Most significantly (and most paradoxically from the idealist standpoint) Stcherbatsky noted that Nāgārjuna denied that there was a firm and ultimate disjunction between the empirical world and the Absolute (between *saṃsāra* and *nirvāṇa*). Because it was stated (in Chapter XXV of the *Mādhyamikakārikā)* that, from the standpoint of absolute truth (*paramārtha-satya*), they are actually one and the same, it might have appeared that Nāgārjuna was denying the idealist's basic assumption of the appearance/reality distinction. But Stcherbatsky saw it another way—and this was his crucial innovation; he stated that Nāgārjuna, like the other Buddhist philosophers he had already explicated, must be arguing that the world of appearance is a false reality. It is only the other world, *nirvāṇa,* that is real. And if, as Nāgārjuna claimed throughout much of the *Mādhyamikakārikā,* the apparent realities of multiplicity, causation, motion, and change are false mental products (note the similarity to his analysis of Dignāga's epistemology), then the "real" world must be, in contrast, an unchanging, uncaused, undifferentiated state of being much like Śaṃkara's *Brahman.* He did not even entertain the idea that, by denying any distinction between *saṃsāra* and *nirvāṇa,* Nāgārjuna might be specifically denying that there is a disjunction between separate worlds of appearance and reality. His presupposition, as a thorough Kantian, was that the disjunction always held; it could not be denied. It is revealing that Stcherbatsky chose to translate only two chapters of

Nāgārjuna's important text, leaving out much of what we now tend to see as the complexity and perspectivism of Nāgārjuna's position. He included Chapter one, which begins with a denial of the existence of all phenomenal entities in Nāgārjuna's characteristic rhetorical form, the *catuṣkoti,* or "four-cornered negation":

> There absolutely are no things,
> Nowhere and none, that arise (anew),
> Neither out of themselves, nor out of non-self,
> Nor out of both, nor at random.
> (Chapter I, verse 1)

And he included a translation of the paradoxical Chapter XXV, which contained, for Stcherbatsky, what might have been the most potentially troublesome of Mādhyamika assertions, but what he chose to read as the clearest Nāgārjunian expression of the Kantian position:

> There is no difference at all
> Between *Nirvāṇa* and *Saṃsāra,*
> There is no difference at all
> Between *Saṃsāra* and *Nirvāṇa.*
>
> What makes the limit of *Nirvāṇa*
> Is also then the limit of *Saṃsāra.*
> Between the two we cannot find
> The slightest shade of difference.
> (Chapter XXV, verses 19 and 20)

Together, these two chapters, along with the translation of Candrakīrti's *Prasannapadā,* were meant to emphasize the point that Nāgārjuna was an idealist and a monist rather than a radical skeptic or a nihilist. Nāgārjuna, Stcherbatsky pointed out, denied the apparent reality of multiplicity, change, and creation, while simultaneously asserting the reality of an ultimate existence. To an early twentieth-century idealist like Stcherbatsky, that meant that the noumenal realm must be one of stillness and undifferentiated stasis: if the phenomenal world were composed of the ceaseless motion of individual particles, then the noumenal world must, by contrast, be a single "motionless whole." Like the European scholars before him, almost all of his attention was focused on the difficulties of the relation between *saṃsāra* and *nirvāṇa.* But unlike his predecessors, he read Nāgārjuna's seemingly paradoxical assertion of the identity of two levels of reality as an affirmation of Kant's phenomenal–noumenal relation. His was not an illogical denial of reality.

Stcherbatsky's chapter on European parallels to the Mādhyamika position most clearly exposes his attempt to reformulate Nāgārjuna's thought as a brand of European idealism. In this section of his writings he originated a scholarly tradition that continues up to the present time: to use currently popular European figures (in this case, F. H. Bradley and Hegel) and to identify Nāgārjuna with them.

> Very remarkable are then the coincidences between Nāgārjuna's negativism and the condemnation by Mr. Bradley of almost every conception of the every day world, things and qualities, relations, space and time, change, causation, motion, the self. From the Indian standpoint Bradley can be characterized as a genuine Mādhyamika. But above all this parallelism we may perhaps find a still greater family likeness between the dialectical method of Hegel and Nāgārjuna's dialectics. . . . Both philosophers assure us that negativity (*śūnyatā*) is the soul of the universe, *"Negativitat ist die Seele der Welt."*[40]

It was in his discussion of parallels that he took his most extreme position on the identity of Mādhyamika and Advaita Vedānta:

> To characterize Nāgārjuna as a "nihilist", means to make a misleading comparison, since his condemnation of logic is only one part, and not the principal one, of his philosophy. . . . Upon the Indian side we must first of all point to the almost absolute identity with Vedānta . . . it follows from this identity that all the points of contact which Prof. Deussen has really found, or imagined to have found, between Schopenhauer and Vedānta, will equally apply to Nāgārjuna.[41]

The Kantian revaluation of Mādhyamika Buddhism was, explicitly, an attempt to exonerate Nāgārjuna from the charge of nihilism. Stcherbatsky was in agreement with other Buddhologists like Burnouf, Schayer, and even his rival, La Vallée Poussin, that a philosopher could be taken seriously only if he held that some entity, term, or being was "real" in an absolute sense. In his view, and in the view of many of his contemporaries, a philosophy was intelligible only to the extent that it examined the relation between apparent reality and the transcendental absolute. For Stcherbatsky, Mādhyamika met these idealist qualifications, and met them in a particularly intriguing way. Nāgārjuna's thought was, for him, like the Abhidharma and the Buddhist logicians, a variation on the ubiquitous theme of appearance versus reality. *Śūnyatā* meant "relativity": the unreality of interdependent "dharmas" or "point-instants," as they exist in themselves. As such, the Mādhyamika view of ultimate reality was radically different from the pluralisms of the early and late Buddhist schools. But, Stcherbatsky claimed, this position should not be construed to deny the existence of an absolute in Nāgārjuna's philosophy. Just as in his exposition of Dignāga's and Dharmakīrti's philosophy in *Buddhist Logic,* Stcherbatsky interpreted Nāgārjuna's thought in *The Conception of Buddhist Nirvāṇa* as a description of the nature of absolute reality (albeit a negative one). Most importantly, Stcherbatsky repeated his claim that Buddhist philosophy in all of its forms was focused on man's inability to conceptualize ultimacy. Sarvāstivādins, Yogācāras, and Mādhyamikas were all engaged in a similar effort to describe the disjunction between "concepts" and "intuitions" or "appearance" and "reality." In his view, the Mādhyamika texts, as opaque as they might seem at first, exhibited the basic concerns of all authentic philosophies. Nāgārjuna's paired terms were *nirvāṇa/saṃsāra* and *paramārtha-satya/saṃvṛtti-satya.* Stcherbatsky recognized them at once, and retranslated them as *noumena* and *phenomena.* In this way, through Stcherbatsky, Kant's Copernican strategy brought about a revolutionary revaluation of Buddhist philosophy among European scholars. Like the Vedānta, Buddhist thought, even the Mādhyamika, could now be read as a precursor to Western idealist philosophy. To Europe in the early twentieth century, this was an automatic conferral of respectability.

The Indianization of Stcherbatsky

The surest indication that Stcherbatsky's radical restatement of Mādhyamika philosophy had become the new standard reading was the emergence of T. R. V. Murti's *The Central Philosophy of Buddhism* (1955) as a basic text in Buddhist studies. Murti had been trained in Western philosophy and his work bears the unmistakable stamp of Stcherbatsky's influence. The preface opens with a quote from *The Conception of Buddhist Nirvāṇa* and Murti, like Stcherbatsky, selects as his adversaries those who dismiss Nāgārjuna as a nihilist. (He refers to the nihilism interpretation as "uninformed" and "unjustified."[42]) Murti labels Mādhyamika as absolutist rather than nihilist and says that the introduction of Mādhyamika into Buddhist thought brought about

> a revolution from a radical pluralism (Theory of Elements, *dharma-vāda)* to an as radical absolutism (*advaya-vāda*). The change was

> from a plurality of discrete ultimate entities (*dharmāḥ*) to the essential unity underlying them (*dharmatā*).[43]

Like Stcherbatsky, Murti follows the historical schema of the "Three Swingings of the Wheel of the Law." But more substantively, he accepts Stcherbatsky's suggestion that Mādhyamika introduced into Buddhist philosophy the idea of an underlying unity, a transcendental absolute. For Murti, the Mādhyamika reality is, like Śaṃkara's *Brahman*, an undifferentiated unity. He assumes, like his predecessor, that absolutism (the opposite of nihilism, for these gentlemen) is identical with monism.

There is more complexity to Murti's examination of Mādhyamika than there is to Stcherbatsky's and, significantly, far more evidence of inconsistency. Consequently, as Karl Potter stated, "Murti seems to waver"[44] in his overall analysis of Nāgārjuna's thought. He says that Nāgārjuna's philosophical perspective is one that denies interest in metaphysics and condemns speculation about the nature of ultimate reality. He repeatedly emphasizes that Nāgārjuna neither advances a speculative thesis of his own nor intends to offer anything beyond criticism of other philosophical doctrines:

> Philosophy, for the Mādhyamika, is not an explanation of things through conceptual patterns. That is the way of dogmatic speculation (*dṛṣṭi*); but this does not give us the truth. The Dialectic is intended as an effective antidote for this dogmatic procedure of reason; it is the criticism of theories (*śūnyatā sarva-dṛṣṭinam*) . . . criticism of theories is not another theory; *śūnyatā* of *dṛṣṭis* is not one more *dṛṣṭi*, but is *prajñā*—their reflective awareness.[45]

Yet, while insisting that the purpose of the Mādhyamika dialectic is only to unmask the absurdities of other positions, Murti repeatedly states that Mādhyamika seeks to reveal the essential nature of *śūnyatā*—to uncover the fact that "the Absolute (*Śūnya*) is the universal, impersonal reality of the world."[46]

In fact, Murti claims, Mādhyamika absolutism, in its basic outlines, is very similar to the systems of Vedānta and Vijñānavāda (Yogācāra):

> In all these systems, the Absolute is transcendent, totally devoid of empirical determinations. The Absolute is immanent too, being the reality of appearance. The Absolute is but the *phenomena* in their essential form . . . Every absolutism is really an *advaita* or *advayavāda*, non-dualism; they do not establish the absolute, but just reject duality as illusion.[47]

Throughout his book, Murti links what he calls the "three absolutisms of Indian philosophy:" Vedānta, Vijñānavāda, and Mādhyamika,[48] and notes their commonalities. But, unlike Stcherbatsky, who argued that Mādhyamika and Vedānta were essentially identical, Murti carefully attempts to differentiate between the three schools. And it is when he discusses their differences that he reveals preconceptions about what any philosophical school must try to do, and why he sees Nāgārjuna as playing Kant to Śaṃkara's Hegel. Murti admits that he has trouble in understanding the relation between the realms of absolute and phenomenal existence in Mādhyamika[49] and that he is more comfortable with the other absolutist outlooks. In both Vedānta and Vijñānavāda, illusion is seen to appear on a real ground (*adiṣṭhāna*) that makes that illusion possible. *Brahman* and *vijñāna* both act as causal substrata for phenomenal existence. But Nāgārjuna denies entirely the idea of a causal substratum and, consequently, Murti's efforts to characterize Nāgārjuna as an absolutist are not as easy as he would like:

> It is not true to say, as is done by Vedānta and Vijñānavāda, that the Mādhyamika conceives illusion to occur without any underlying ground (*niradhiṣṭhānabhrama*). *Tattva* as *Dharmatā* or *Bhūtakoṭi* is accepted by the Mādhyamika as the underlying ground of *phenomena*. But it is now shown by him to be immanent in experience, how *Dharmatā* activates and illumines empirical things. Not that the Mādhyamika takes the Absolute and the world of *phenomena* as two different sets of entities; but the Absolute is nowhere explicitly shown to be *in* things constituting their very soul. The relation between the two is not made abundantly clear. This may be said to constitute a drawback in the Mādhyamika conception of the Absolute.[50]

Murti says that it is a drawback in the Mādhyamika presentation that there is no description of how the realm of absolute truth upholds and acts as a ground for the realm of phenomenal truth. He is troubled by the fact that "the Absolute is nowhere explicitly shown to be in things constituting their very soul." What he does not make clear, however, is why he considers this "grounding" to be necessary for the Mādhyamika perspective, or why the lack of such a discussion in Nāgārjuna's text should be interpreted as a weakness. Murti is satisfied with what he sees as Nāgārjuna's view of the relation between the absolute and the conditioned on the "epistemological level," but he finds that this conflicts with a lack of theoretical relation on the "causal" or "metaphysical" level. He says that Nāgārjuna denies any interest in metaphysics—and yet, paradoxically, Murti requires a discussion of causal and ontological relations.

In an article written some years after the publication of *The Central Philosophy of Buddhism,* Murti attempts to clarify his position on this problem by introducing a distinction between "epistemological" and "metaphysical" absolutisms. Murti states that, in epistemological terms, *paramārtha-satya* is the knowledge of the real "as it is without any distortion (*akṛtrinam vasturūpam*)."[51] This distortion is the result of the process of conceptual thought, the process whereby the human mind makes discriminatory judgments. It is only through the final abolition of *saṃvṛti* that knowledge of *paramārtha* can be attained: through the elimination of discursive, logical ratiocination, true insight into "the Real" is possible. Thus, in Nāgārjuna's system, "*saṃvṛti* is the means and *paramārtha* is the end."[52]

When he speaks in metaphysical terms, however, Murti will insist that Mādhyamika affirms the reality of an ontological substratum—a monistic *paramārtha* that is obscured by the conceptual veil of *saṃvṛti*. The Mādhyamika view, he says, is that reality cannot be clothed in "empirical terms and concepts" (*prapañca, vikalpa*).[53] But it is not the Mādhyamika view, he insists, that the Absolute itself be considered unreal. Murti says the belief that Emptiness itself is "empty" would lead only to infinite regress. So he asserts that Nāgārjuna's description of reality as *śūnya* is not itself empty, but is an intuitive perception of the Absolute. Nāgārjuna's *śūnyavāda* is seen as affirming a universal substratum, the *paramārtha,* which exists behind the illusory curtain of *saṃvṛti* and which is intuited through *prajñā.* By stating the Mādhyamika position this way, however, Murti realizes that, like Stcherbatsky, he is virtually equating *śūnyatā* with the Vedānta notion of *Brahman*—the all-encompassing, unqualifiable absolute. And this, significantly, is where he clearly parts ways with his European mentor. Murti is himself a devout Hindu and consequently needs, for personal reasons that Stcherbatsky did not have, a way to avoid this equivalency. Murti states that the two philosophical schools must not be understood as Buddhist and Hindu versions of the same view: "The differences between them are deep and pervasive."[54] So he resorts to the strategy of differentiating between two types of absolutisms: the metaphysical (Vedānta) and the epistemological (Mādhyamika). These two types of absolutism are alleged to be different in methodology but, in Murti's view, they are alike in attempting to uncover the essential "Pure Reality" behind the world of mere appearance.

By claiming that "all absolutisms need not be of the Vedāntic type,"[55] Murti attempts to retain the label of absolutism while avoiding the troublesome identification with Vedānta:

> The Mādhyamika is epistemological in his procedure while the Vedānta is ontological. The former is an *advaya-vāda* (no two views), while the latter is an *advaita* (no two things).[56]

Murti states that the Mādhyamika absolutism offers no definitive positions on questions of ontology but posits distinctions only between categories of ideas and to some extent adjudicates knowledge claims. In this way, Murti claims, Nāgārjuna brought about a "Copernican revolution" in Indian philosophy:

> The position occupied by the Mādhyamika in Indian philosophy is similar to that of Kant in modern European philosophy . . . both the Mādhyamika and Kant can justly be credited with having initiated the critical phase in philosophy in their respective spheres.[57]

By comparing Nāgārjuna with Kant, Murti believes that he has found a way to avoid describing Mādhyamika as a metaphysical absolutism. But it still leaves him with the following problem: what is the nature of the absolute in this "epistemological" system? In what way is it an absolutism? Nāgārjuna does not posit any forms or categories of mind that are to be considered ultimately real, nor does he posit any other candidates for the role of an epistemic Pure Reality. Murti's strategy—converting metaphysics to epistemology—still leaves him uneasy.

In his discussions of the common aspects of all absolutisms, Murti states that the Mādhyamika, Vijñānavāda, and Vedānta schools are all "formally" similar. They differ only in regard to their rhetoric, dialectical methodology, and "possibly with regard to that entity with which they identify the absolute."[58] Moreover, he states that all absolutisms are ultimately reducible to the Mādhyamika position of "No-position:"

> The difference pertains to the mode of approach, to the standpoint from which these systems reach the Absolute, like the center of a circle which is reached from the periphery by different radii.[59]

In his effort to discredit the label of nihilism he argues, like Stcherbatsky, that the Mādhyamika philosophy is centered around the intuition of a transcendental reality, just as the other "Indian absolutisms" center around their particular notions of the "ultimately real." Nāgārjuna's *paramārtha* is different from Śaṃkara's *Brahman* because it is reached by an "epistemological" approach. But it is an "absolute" nonetheless. Inconveniently for Murti, Nāgārjuna does not posit any category of ultimate existence, epistemological or metaphysical, that could be said to act as a ground for *saṃsāra*. Terms such as "consciousness" "mind," "spirit," "idea," "understanding," or "sensation," are not available to act as possible epistemic realities. There is only *śūnyatā*. As a result, Murti vacillates between his arguments that Nāgārjuna is primarily concerned with revealing a transcendental pure reality and his demonstrations that Nāgārjuna is primarily concerned with revealing a transcendental pure reality and his demonstrations that Nāgārjuna does nothing beyond offering critiques of other philosophical positions.

Murti argues that Nāgārjuna believes that the absolute transcends discursive thought—that it is unreachable through the use of rational judgments. empirical investigation, or philosophical speculation. He also argues, again, that Nāgārjuna must not be construed as a nihilist:

> The dialectic should not be taken, as is done by the uninformed, as the denial of the Real—Nihilism.[60]

He is adamant that the Mādhyamika's "no-doctrine-about-the-real" attitude should not be seen as a "no-reality" doctrine. It has been the opinion of most scholars that he succeeded in mounting persuasive demonstrations of both of these points. I am arguing, however, that he succeeds only at the cost of interpreting Nāgārjuna's thought as an "absolutism" that hides thinly veiled metaphysical commitments behind the camouflage of "empistemological" rhetoric. The substitution of terms hardly changes the fact that, by Murti's reading (as by that of Stcherbatsky before him), Nāgārjuna's basic philosophical concerns are taken to be identical to those of Śaṃkara. Murti may insist that the terminology and surface methodology are different, but, he thinks, it is the relation of the unreal, phenomenal world of *saṃsāra* to the real, noumenal world of *nirvāṇa* that is crucial to the Mādhyamika philosopher as well as to the Vedāntin. The strictly epistemological terminology does little beyond avoid the unwanted terminology of metaphysics that was prevalent in earlier idealist interpretations. It is true that Murti, unlike Stcherbatsky, tries to avoid characterizing the nature of the absolute reality. (In a similar way, Kant denies that the thing-in-itself is available to our direct inspection of description.) But Murti, like Kant, must be committed to the absolute reality of both the thing-in-itself and the notion of the constituting capacity of the mind. As a result of his inability to identify a similar dual commitment in Nāgārjuna's text, the tension in Murti's reading is left unresolved. Ironically, Murti's determination to characterize Nāgārjuna as an absolutist, and thus his inability to find a way between unsatisfactory choices in interpretation, may be the result of having taken the charge of nihilism too seriously. But he could hardly see it otherwise. For an idealist such as Murti, the choice is necessarily between reality and un-reality, and it is a philosophical given that the world as ordinarily perceived is erroneously taken as real. He sees Nāgārjuna's dialectic as effectively demonstrating the "unreality" of phenomenal entities and relations. Therefore, to avoid conceding that Nāgārjuna was a nihilist, he feels it is necessary to argue that Nāgārjuna believes there is an ultimately real noumenal realm behind the shadow world of appearance.

Stcherbatsky's Kantian reading, as reinterpreted by Murti, had become so influential that it was now the standard view for both European and Indian scholars. Through the export of German idealism, European imperialism had thus triumphed in the philosophical as well as the economic and political arenas. But European philosophical assumptions were dramatically changing. Changes in the understanding of the Indian philosophical texts were unavoidably to follow.

3

Analytic India

The Logic of Own-Being

The Idealists had established Nāgārjuna's reputation. By the time the philosophical influence of G. E. Moore, Bertrand Russell, A. J. Ayer, and the Vienna Circle was felt by students of Asian philosophy, Nāgārjuna's prestige had risen sharply. After the debates between Stcherbatsky and La Vallée Poussin in the late 1920s about the nature of the Buddhist *nirvāṇa,* and after Stcherbatsky had demonstrated the Kantian nature of Nāgārjuna's treatise, few scholars were willing to dismiss Mādhyamika as irrelevant. Murti's contribution, some 25 years later, reinforced the idea that Nāgārjuna was "central" to the entire structure of Buddhist thought. It had become a fact. But younger contemporaries were finding a new (nonidealist) way to write about Nāgārjuna, as their training enabled them to do—or required them to do. Nāgārjuna did not claim, they said, that the conventional world is "unreal," or less real than the absolute. He claimed only that it is "empty"—devoid of *svabhāva* (variously translated as "own-being" or "self-existence"). Seen in this way, there is no question of determining the ultimate reality of either *saṃvṛti* or *paramārthsatya.* Conventional reality is just as "real"—just as "empty"—as absolute reality and, thus, as Nāgārjuna contended, it is identical with it. According to this new interpretation, Nāgārjuna intended to show that *svabhāva* is internally inconsistent, that the world cannot be shown to contain "intrinsically real" entities or relations. Consequently, some scholars began to argue that Nāgārjuna claimed neither that the world is unreal nor that it is bifurcated into two levels, the real and the apparent. For these younger writers, influenced by the scientific approach of logical positivism and the rigorous feel of analytic philosophy, Nāgārjuna was a dialectician: he constructed carefully turned arguments and counterarguments to defeat philosophical opponents. Beginning in the 1950s, these scholars concentrated less on "mystic intuition of the Saint" than on the "logical analysis" of Nāgārjuna's verses. They produced (and this was their intention) a body of interpretation that was more in keeping with the kind of work being done in Anglo-American departments of philosophy.[1]

The distinction between the new analytic interpreters and their idealist predecessors was, in large part, stylistic. The visual appearance of their writings differs radically. Charts, graphs, formulas, symbols, and equations decorate the pages of these new Asianists, when they tackled those subjects they deemed properly rigorous (Nāgārjuna's dialectical strategies, Buddhist "theories of meaning," Hindu "logical methods"). The older poetic comparisons of Nāgārjunian verses with lines from Kant and Hegel were no longer considered truly philosophical, or even reputable scholarship. Instead, the rage to quantify, logically analyze, and symbolically restate the teachings of Nāgārjuna was now overwhelming. The identification of the conditionals, quantifiers, negations, and conclusions in the translation of the *Mādhyamikakārikā* became the occasion for the symbolic restatement and detailed analysis of the logical structure of the Sanskrit sentence. And, predictably, the *catuṣkoti,* or "four-cornered negation," often used by Nāgārjuna to deny four exhaustive alternatives, became the center of attention for a number of analytically oriented Buddhologists.[2] This rhetorical device ("Everything is either true, or not true, or both true and not true, or neither true nor not true; that is the Buddha's teaching." Chapter XVIII, verse 8) had been noted by earlier writers such as Stcherbatsky and Murti, but had never received microscopic attention. In this new atmosphere specific sections of text were selected for detailed analysis and criticism in a way that was foreign to the idealists who had been searching for a unified system of philosophy. The *Mādhyamikakārikā* was no longer considered a monolithic entity to be read and understood in its entirety. It was now a collection of propositions, syllogisms, and logical devices that could be isolated, translated into artificial languages, and judged for their logical validity. This change in style and subject matter indicated far more than a shift in interpretive emphasis, it indicated an entirely new way of reading the text. The modern European interpreter and the author of the ancient Sanskrit work were united by the rules of a universal language: the imperatives of systematic reason, as defined by the modern reader.

One of the most influential and insistent of the analytic scholars, Richard H. Robinson, entered the field in 1957 with an article entitled "Some Logical Aspects of Nāgārjuna's System." This began with a brief history of modern Nāgārjuna scholarship and traced Western interest in Mādhyamika from the initial work done by Burnouf through the twentieth-century interpreters (Stcherbatsky, La Vallée Poussin, Schayer, Murti, and J. W. De Jong), and it concluded with a scathing appraisal of their interpretive efforts. Robinson found their work philosophically unsophisticated and he relegated them to what he termed the "metaphysical" phase of scholarship. They were guilty of having led Western readers into a quagmire of interpretive confusion:

> The "metaphysical" debate has exhibited extreme variety of opinion, and the attempt to describe Mādyamika as an "ism" has led various people to call it nihilism, negativism, monism, relativism, irrationalism, criticism, and absolutism. Attempts to find transformulations based on analogy with Western thinkers have not gone very far. The most usual comparisons, those with Kant and Hegel, are not apposite, because Kant's and Hegel's structures differ too radically from any of the Indian systems in question.[3]

In one paragraph he swept aside the standard scholarly procedures of his precursors: to characterize Indian philosophies according to standard Western philosophical types (or "isms") and to compare Indian philosophers with familiar Western figures. In his view, earlier studies of Mādhyamika were deficient because

> Writers have generally taken the Mādhyamika statements out of context and reshuffled them into some modern pattern, with the result that the intrinsic structure of argumentation has not been clearly discerned. An even deeper weakness of the metaphysical approach is that it seeks to answer our questions, rather than to identify Nāgārjuna's questions.[4]

Rather than idealist philosophical speculation, Robinson suggested, the model for interpretive studies of Buddhist texts should be contemporary linguistic science, "where more rigorous analysis and more closely controlled reasoning have captured the field."[5] And above all, the "one topic which particularly needs separate and detailed inquiry is the role of logic in Mādhyamika."[6] Stcherbatsky's and Murti's discussions of logic had been insufficiently rigorous, he stated; specialists in modern logical analysis were needed who could clear up the confusion created by too much metaphysical speculation. Only analytic clarity would determine if "Nāgārjuna was actually saying something meaningful" and if Nāgārjuna's arguments could stand up to modern methodology.[7] Robinson cited Stanislaw Schayer and Hajime Nakamura as pioneers in the attempt to apply symbolic logic to Indian philosophy and asserted the "superiority of modern scientific, notational logic as an instrument for investigating Indian logic. Notational statement," he claimed, "avoids the pitfalls and awkwardness of linguistic statement and rhetorical logic."[8] Mathematical symbols avoid the awkwardness of language.

With his self-avowed predilection for symbolic logic expressed in the opening pages, Robinson carried out his program by rewriting as many phrases from the *Mādhyamikakārikā* as possible in as many different notational systems as he could muster. Equations, tables, and mathematical symbols filled the pages and instead of a general description of Nāgārjuna's thought he began his examination of the *Mādhyamikakārikā* with a section entitled "Basic Theorems and Rules of Inference." To produce numbered lists of examples as evidence, Robinson carefully examined the text for Nāgārjunian illustrations of such Western logical mainstays as "the principle of contradiction," "the law of the excluded middle," and "the law of identity." He identified the syllogistic forms of *modus ponens* ("p implies q; p, therefore q") and *modus tollens* ("p implies q; not q; therefore not p") among Nāgārjuna's hypotheticals, located instances of logical fallacies among Nāgārjuna's arguments, and offered redescriptions of the *catuṣkoti* form of negation in several symbolic and diagrammatic forms:

1. [Ax v −Ax v Ax . −Ax v −(Ax) . − −(Ax)]
2. "a," "−a," "a−a," "−(a−a)"
3. ab = 0
   ab = 0
   ab ≠ 0 . ab ≠ 0
   ab = 0 . ab = 0[9]

The article went on to list brief definitions of key terms like *nirvāṇa, svabhāva,* and *pratītyasamutpāda* but, instead of attempting, like his predecessors, to devote his energy to a discursive explication of these concepts, Robinson concentrated almost entirely on what he called Nāgārjuna's "dialectical apparatus." The remainder of the article included individual and highly schematized sections on "Negation," "Quantification," "The Tetralemma" (the *catuṣkoti*), and "Dilemmas." His general principles were to avoid generalization about the underlying meaning of the Mādhyamika system, to avoid both the vagaries of common language and the romanticism of metaphysical language, and to avoid indulgence in identification between Nāgārjuna and any Western philosopher. The Russellian notion that linguistic utterances can be converted to a pure, universal, logical structure had finally come to Buddhist studies, and with it, the rejection (as merely "intuitive") of the earlier Kantian and Hegelian readings of Nāgārjuna's thought.

Robinson wrote like a Vienna Circle positivist; for example, he said that descriptive terms are irrelevant,

> except as signalling emotional acceptance or rejection. Nor are we further ahead for the knowledge that emptiness is transcendental, or nihilist, unless we know the configuration of qualifications with which such concepts have meaning.[10]

His article was the first major polemic in what was soon to be a barrage of attacks against the old style of reading Indian philosophical texts. It concluded with the accurate prediction that the logical analysis of Nāgārjuna and other Indian philosophers had only just begun:

> This article has made relatively slight use of the resources of modern logic, but it is possible to transcribe the *Kārikās* entirely, chapter by chapter, into logical notation, thus bringing to light formal features which do not appear from the consideration of examples taken out of context and listed topically. In short, the logical analysis of Nāgārjuna is far from complete.[11]

From the standpoint of the "Age of Analysis," the Kantian and Hegelian parallels presented by the pioneers of Mādhyamika studies had been of little use except to obstruct an understanding and appreciation of the true import of Nāgārjuna's reasoning. Truth in 1957 was the province of logicians.

For Robinson, modern logical analysis was not only the key for unlocking the hidden meaning of ancient texts, but also a reliable tool for criticizing the validity of ancient arguments. Throughout his writings on Mādhyamika, Robinson insisted that there was a "standard strategy of refutation."[12] This strategy, according to Robinson, was used to negate all possible wrong views in one comprehensive schema and was built around the single most important principle in the Mādhyamika logical system: the concept of *svabhāva*. It was on the merits of Nāgārjuna's treatment of *svabhāva,* he contended, that the success or failure of the entire Mādhyamika could be judged, and it was Nāgārjuna's use of the principle of *svabhāva* from which the uniqueness of the Mādhyamika system derived. All other Indian philosophies, Robinson argued, posited one or more entities on which other entities depend but which themselves are independent. These are the independent realities that were said by the various systems of philosophy to act as a causal or cognitive substrata for the universe of observable reality. Advaitins held that all things depend on *Brahman* but that *Brahman* is wholly uncaused and independent. Sāṃkhya philosophy contended that the manifested constituents that make up the empirical world depend on the unmanifested *prakṛti,* but not vice versa. Nyāya-Vaiśeṣikas and Mīmāṃsākas posited discrete atomic substances that do not depend on any other entities and on which all other substances and all qualities depend. Within the Buddhist tradition, Abhidharmists posited the independent existence of the elemental *dharmas,* Yogācāra philosophy held that all things depend on *vijñāna,* and the "logical" school of Dignāga and Dharmakīrti imputed ultimate reality to a substratum of atomic point-instants. All these ultimate substances, Robinson suggested, exemplified what Nāgārjuna meant by *svabhāva.* And although Nāgārjuna did not argue explicitly against the views of all these schools (he historically preceded a fair number of them), Robinson argued that Nāgārjuna attempted, through a single refutational strategy, to anticipate any philosophical position that might assume the possibility of existential ultimacy. It was this strategy that Robinson was determined to isolate and critique in *Early Mādhyamika in India and China* (1965) and in his diagrammatic overview of Buddhist and Hindu thought, "Classical Indian Philosophy" (1967).

Robinson claimed that Nāgārjuna, in his attempt to negate the basic assumptions of all other philosophical schools, defined *svabhāva* for his own purposes as a self-contradictory idea. Consequently, Nāgārjuna may have been guilty of battling dragons of his own creation. According to Robinson's reading of the *Mādhyamikakārikā,* no entity or state can be said to possess "own-being" or "intrinsic self-existence" and at the same time to exist in a relational context with other entities or states. *Svabhāva,* it was stated, "refers to something which cannot be made and which has no mutual correspondence with something else."[13] Thus, the *Brahman/Ātman* of the Vedānta or the nine substances (*dravya*) of the Vaiśeṣika cannot be both utterly independent of causality and existentially involved in it. Robinson said that, for Nāgārjuna, self-existence and relational existence were mutually exclusive. And since it can be demonstrated that all things in the world are to some extent interdependent—and if, as Nāgārjuna contended, ontological dependence on something else entails a lack of "own-being"—then, necessarily, all normally existent things are without intrinsic self-existence. Nāgārjuna's technique, according to Robinson, was to repeatedly prove that *svabhāva* is a logical impossibility in a world constructed of complex interrelations:

> If an entity is real, it has own-being and cannot change; if an entity is unreal, it has no own-being which might change. Own-being is changeless, and dependent co-arising manifests a semblance of change. This fundamental incompatibility is the anvil on which Nāgārjuna hammers all his opponents' propositions.[14]

The following pattern, conceived by Robinson and expanded here, represents Nāgārjuna's consistent negational strategy[15]:

1. The opponent states that C relates A and B. (i.e., C is a quality that A and B have in common, a principle that affects both A and B.)
   [C > A; B > C ∴ B > A]
2. A and B must be either identical or completely different.
   a. If they are identical, then C is not applicable, since it requires two different terms to bring into relation.
   b. If they are completely different, C is again inapplicable, since A and B can have no qualities in common and so cannot possibly be related.
3. Therefore, it is false that C relates A to B.

Robinson contended that the requirement that A and B must be either completely identical or completely different, rather than partially related, followed from the self-contradictory definition of *svabhāva* as existent and yet "not dependent." Limited or qualified relations were excluded from Nāgārjuna's discussion because he was exclusively concerned with problems of "essence":

> What pertains to part of an essence must pertain to the whole essence. A defining property is either essential or non-essential. If it is non-essential it is not really a defining property of an essence. If it is essential, then the essence can never be devoid of the property.[16]

For Robinson, Nāgārjuna's seemingly paradoxical definition of *svabhāva* was no more than an attempt to trap his philosophical opponents—an attempt that Robinson found ultimately unsuccessful. "The validity and relevance of Nāgārjuna's refutations hinge upon whether his opponents really upheld the existence of own-beings (*svabhāva*) as he defined the term."[17]

Robinson tried to demonstrate that some of the assumptions that Nāgārjuna attacked were not really the assumptions that many of his opponents held. For example, Nāgārjuna assumed that all extended entities are seen as divisible and composite, and so they are necessarily impermanent and unreal. An indivisible entity could have no extension. But Robinson stated that several schools—Vaiśeṣika and Sarvāstivāda, as well as some that borrow from Upaniṣadic doctrine—referred to *ākāśa* (ether, space) as omnipresent and indivisible. Thus, the idea of indivisible, extended entities was available to his opponents. Likewise, Nāgārjuna contended that all existent entities are "arisen," that there is no unmanifested existence. Sāṃkhya philosophy, however, posited the existence of unmanifested *prakṛti,* things "existing" in another state before they arise and after they cease. In addition, said Robinson, Nāgārjuna assumed that any ultimately real entity would have to be a totally simple unity without any diversity. But again, the Sāṃkhya school did not actually make this assumption, as *prakṛti* was said to consist in its ultimate state of three *guṇas* (strands) and, therefore, to exist intrinsically in diversity.[18] As an overall assessment, Robinson wrote that

> This is enough to show that Nāgārjuna did not achieve the ambitious task that he set himself. One reason is that own-being (*svabhāva*) as he uses it is too general a concept. He lumps together too many quite disparate things, and to make them fit he has to exclude some essential features of each school's doctrine. If he had succeeded in refuting all philosophical views, it would have been a momentous development in the history of philosophy . . . his polemic campaigns were not very successful in achieving their main objective . . . his school did not endanger the existence of its rivals.[19]

According to Robinson, Nāgārjuna's method (trying to demonstrate inconsistency in the notion of *svabhāva*) was insufficient to refute all opposing philosophical views. Many of the competing positions were not actually characterized by the concept of "own-being" as defined for his own purposes by Nāgārjuna. In Robinson's estimation, therefore, Nāgārjuna's logical program failed because of lack of modern analytical expertise. Nāgārjuna ultimately failed to defeat his opponents because his notion of *svabhāva* was dialectically ineffective. As a result, many of his opponents were allowed to escape.

Unlike his idealist predecessors, Robinson adopted a negative stance toward Nāgārjuna's text, an interpretation that nevertheless allowed him to consider Mādhyamika Buddhism seriously. Stcherbatsky's primary concern was to combat the view that Mādhyamika was only nihilistic nonsense. Murti, in turn, had insisted that Nāgārjuna was among the most important of Indian thinkers, and at least equal to the great religious and philosophical minds of the West. But Robinson, who claimed to be interested in nothing beyond the objective examination of logical method, approached the text with what he imagined to be clinical disinterest. In his view, Nāgārjuna's aim was the logical refutation of all competing philosophical positions. And, because of what Robinson discerned as the inevitable ineffectiveness of the Mādhyamika "standard strategy of refutation," he portrayed Nāgārjuna's position in the history of philosophy as that of an ambitious but ultimately unsuccessful dialectician.

Robinson's penchant for notational restatement exemplifies the analytic aesthetic; his pronouncement on Nāgārjuna's failure tells us much about his philosophical morality. For a scholar steeped in the ubiquitous positivism of the 1930s and 1940s, philosophical statements were significant only to the extent that they could be hardened into notated shorthand and then checked for logical validity and dialectical effectiveness. The goal of a philosophical assertion would be to refute an opponent or to establish an unassailable position. Philosophical schools were armies with arsenals of logical weapons that could be compared and judged for relative effectiveness. Individual philosophers were generals who deployed their school's arguments and who would either win or lose according to the brilliance of their field strategy. In Robinson's own words, Nāgārjuna had failed because "his school did not endanger the existence of its rivals." This military sensibility, which required that mathematical symbols be substituted for words and equations for sentences, also required a victor and a vanquished in any philosophical exchange. There were only winners or losers; no prisoners were taken. Metaphors of violence and competition abound:

> Was he in fact using reason to demolish reason?[20]
>
> On this road, inference, especially the destructive dialectic, plays a crucial part.[21]

> so Nāgārjuna has little trouble demolishing any proposition whose terms are held to have *svabhāvas*.[22]

> He claims that he is not affirming propositions of his own but is just using his opponents' statements to reduce his opponents' theories to absurdity. . . . In fact, Nāgārjuna frames his opponents. His case rests on several assumptions that are not shared by those whom he is attacking.[23]

> His polemic campaigns were not very successful in achieving their main objective.[24]

> Relations are bad, and the ideal is to cut them off.[25]

If Nāgārjuna had been truly successful in his warfare against all other philosophical positions, Robinson maintained, "it would have been a momentous development in the history of philosophy." But that other philosophies and philosophers continued to exist and offer their own view even after Nāgārjuna's attack had been launched was, in itself, proof enough for Robinson that Nāgārjuna was not a successful warrior.

Robinson saw a simple choice between eliminating complexity and paradox from the Mādhyamika system or dismissing Nāgārjuna's text as mystical foolishness. In his desire to choose the former alternative, he reduced the content of the *Mādhyamikakārikā* to the repetition of one dialectical tactic based on the incompatibility of *svabhāva* and existence. And he insisted that all of Nāgārjuna's verses could be explicated without confusion or resort to mystical formulations:

> There do not seem to be any real paradoxes in the *Stanzas*. The seeming paradoxes are easily resolved once the definitions and the fundamental absurdity of the concept of own-being are taken into account.[26]

Robinson wrote in short, schematic paragraphs, offering quick summaries of Nāgārjuna's position in articles that were seldom more than 10 pages in length.[27] These brief explications were meant to simplify the complexities of Nāgārjuna's position and to eliminate any trace of vague, metaphysical speculation. Like the other members of his generation, Robinson did not seek transcendent mysteries. The common desire among the analytic community was for total clarity (through the use of symbolic notation), absolute decisiveness in philosophical disputes (clear winners and losers), and the eradication of paradox and mystery. The ideal for philosophical scholarship was to be found in the mathematical sciences. Brevity, clarity, and decisiveness were the criteria of success. If Mādhyamika philosophy could be explicated, it would have to be done in this manner. Otherwise it was not worthy of consideration from a philosophical standpoint.

As Murti and Stcherbatsky had done for their generation, Robinson succeeded in advancing Nāgārjuna's prestige among the analytically minded scholars of a later *Zeitgeist*. By downplaying Murti's portrayal of Nāgārjuna as a mystic and a philosophical idealist, Robinson actually continued the idealist program of drawing attention to Nāgārjuna as one of the central figures of Asian intellectual history. He did not deny that there was a mystical element in Mādhyamika (he included a short section on the role of mysticism in two of his publications), but he did shift the emphasis from spiritual enlightenment to philosophical rigor. Although Murti and Stcherbatsky had concentrated their attentions on "emptiness" (*śūnyatā*) as the mysterious, central term in the Mādhyamika canon, Robinson turned his analytic scrutiny to the form of the Nāgārjunian dialectic and the role of conceptual analysis. Nāgārjuna, the idealist, had become Nāgārjuna, the logician.

Karma and Causation

Positivism substitutes the aesthetic ideals of clarity, brevity, and decisiveness for the romantic ethos of idealist philosophy. Robinson's determination to convert Nāgārjuna's verses into symbolic equations and to portray Nāgārjuna as an audacious but failed warrior was not the only positivist attempt in the 1950s and 1960s to rework the accepted idealist interpretations of Nāgārjuna's text. The logical analysis of the arguments of the *Mādhyamikakrikā* had become the new standard approach to the subject among younger scholars. But positivist-influenced writers more sympathetic to Nāgārjuna's aims than Robinson found the Buddhist philosopher troublesome when attempting to fit him into a systematic survey of Indian thought. One of the clearest examples of this difficulty can be found in the work of Karl Potter, who, like Robinson, was an American Sanskritist with a fascination for methods more "scientific" than had previously been employed in his field. Like a Skinnerian therapist intent on expunging all mention of emotions, hidden motivations, and the unconscious from psychological discussion, Potter (in *Presuppositions of India's Philosophies*) redescribes the Indian philosophical endeavor as a purely rational game with fixed rules—a test of skill and control based entirely around mechanical laws of causation. Replete with schematic charts, 15 pages of Venn diagrams, and detailed analyses of syllogisms, Potter's book goes even further than Robinson's had in modernizing, masculinizing, and sterilizing the twentieth century's approach to Indian thought (all in accord with the positivist enthusiasm for the rigor of the laboratory). In this system, religion has been turned into something like billiards: a complex set of cold, causal interactions on a limited and predetermined field of play. All that matters is practical success. If you make the right move, you will get the desired result. If you make an error in calculation or execution, you will suffer the consequences. Potter's obsession is with causal chains and final results. Consequently, he sees Indian philosophers as participating in this same obsession.

Potter adopts a single, explanatory theory of human nature as the basis for his study. This bias allows him to create a comprehensive framework in which Nāgārjuna (and all other Indians philosophers) can be placed. But in the process he encounters many of the conceptual difficulties that Western scholars were having with Mādhyamika before Stcherbatsky and Murti had saved its reputation by turning it into German transcendentalism. In his effort to depict Nāgārjuna as one of the rational participants in the intellectual game of Indian philosophy, Potter is forced into a subtle, yet serious, inconsistency. On the one hand, he acknowledges Nāgārjuna's importance (his "centrality," to quote Murti) in Indian intellectual history: by the time Potter had received his education, Nāgārjuna's position as one of the major figures of Asian thought was considered virtually indisputable. On the other hand, however, he is unable to make Nāgārjuna's thought fit neatly into his architectonic schema and consequently has little to say about Mādhyamika. Nāgārjuna receives far less detailed attention in Potter's book than many other more minor figures, and he is often mentioned only in passing. As Nāgārjuna's concerns do not seem to mesh with his own basic theoretical commitments, Potter is reluctant to treat Mādhyamika with a thoroughness consistent with the rest of his study. To understand the reasons for his difficulty, it will be useful to take the title of Potter's work seriously and concentrate on "presuppositions," on what he sees as the basic beliefs of Indian culture. And of course, Potter has his own presuppositions.

Indian philosophy, Potter argues, can be effectively understood without reference to amorphous and nonverifiable factors such as morality, aesthetics, or mysticism. Indian religious traditions, he insists (with a conviction reminiscent of the Victorian horror of heathen immorality), cannot be properly thought of as "moral" in the ordinary Western sense. Power, control, and freedom are the only real values of Indian culture:

> The ultimate value recognized by classical Hinduism in its most sophisticated sources is not morality but freedom, not rational self-control in the interests of the community's welfare but complete control over one's environment—something which includes self-control but also includes control of others and even control of the physical sources of power in the universe.[28]

The protagonists of Indian epics, *yogis,* and *gurus,* Potter argues, are culture heroes because they attain the primary goal of Indian philosophy: freedom from both inner and outer constraint and the power to control oneself and one's world. In Potter's estimation, there are no really significant abstract debates in the Indian tradition about the nature of virtue, as there are in the European literature that traces its ancestry back to Plato. This is, in the Indian context, a practical necessity. "The moral life," or life of *dharma,* is nothing more than one of a series of possible steps on a path to *mokṣa* or release: ultimate freedom and power. Ethical matters are strategic choices designed to further the aims of the participant. In themselves they are of little importance. For Potter, the need for personal liberation seems to stem from the centrality of the doctrine of *karma* in Indian culture. Karmic law—the doctrine that ensures that every individual will inevitably receive just recompense for all morally relevant acts in this or any future life—is described by Potter as the basic philosophical and practical problem for virtually all members of South-Asian cultures. The various forms of Hindu, Buddhist, and Jain religious traditions are, he contends, examples of the possible strategies devised by Indian thinkers to deal with the rules of the karmic game.

Like many previous scholars, Potter is struck by the absence in the Indian tradition of a Platonic/Christian concern with virtue as an abstract ideal.[29] But unlike his preanalytic predecessors he does not view this apparent lack of concern for ethics as an occasion for a call to missionary action. He sees this absence as an indication of great sophistication and modernity on the part of the ancient Indian pandits. Like an entire nation of logical positivists or clinical psychologists, Indians seemed to view life's central problems with an enviable scientific dispassion and emotionless objectivity. If karmically determined transmigration was the trap that must be escaped to avoid countless lives of pain and misery, then, Potter contends, the Indian philosophers were simply intelligent agents who were trying to find the quickest and most effective way out. Their philosophies could be described as either scientific experiments or as serious games played in deadly earnest (and with strict adherence to what a member of the Vienna Circle might call "an empiricist criterion of meaning"). Philosophy is the strategy developed by each player of the karmic game who wants to win the prize: release from an eternity of predetermined lives. Contrary to what was normally said by scholars such as Stcherbatsky or Murti, Indian philosophy need not be viewed as a tradition of abstract speculation and "metaphysical pseudo-statements."[30] It was, perhaps even more than the Western tradition, a scientific attempt to solve real and crucial problems.

But Potter does not stop there. Not only does he describe the spectrum of Indian philosophical schools as a collection of various strategies for winning at the karmic game, he goes on to universalize the idea of *karma,* arguing that all people, Indians and Europeans, are engaged in playing versions of a very similar, and similarly crucial game. Indian philosophy, he contends, is far from an arcane, esoteric tradition. It is a fully practical set of possible answers to a real and contemporary set of questions. And like Schopenhauer before him, Potter is able to enthusiastically assert the timelessness and timeliness of many Indian thinkers, but on his own terms. In Potter's work, these thinkers no longer appear to be anticipating Kant, Fichte, Hegel, Schopenhauer, and Bradley. They now sound like modern Skinnerian psychologists, with theories of behavioral explanation and clinical methodologies. He effects this radical transformation by contending that karmic law is neither an arbitrary presupposition, as it is so often understood, nor even a peculiarly Asian view of the workings of the world. According to Potter, *karma* can be seen as the nonculturally specific human phenomenon we call "habit," and he presents us with a detailed, psychologistic analysis of the role of habits in human life. This rather unexpected psychological emphasis in a book on classical Indian philosophy relies heavily on the schools of positivist learning theory and behavioral description that dominated many social science departments in the 1950s and 1960s. And it is this unacknowledged emphasis that is fundamental to Potter's reading of Nāgārjuna and of all Indian philosophy.

In Potter's view, habits are not insignificant, idiosyncratic, personal traits. They are the overwhelming, internal determinants of the human condition. He argues that to successfully cope with the world, people necessarily generate habitual responses to a great variety of situations. But these habits themselves become problematic as they accumulate, preventing people from dealing creatively and intuitively with new situations by locking them into sets of predetermined responses:

> Habits, necessary to success, constitute a source of bondage. For as one becomes more and more successful through the development of these habitual responses, he tends to become less and less capable of adjusting to fresh or unusual contingencies. Insofar as this hardening of habits does take place, one comes to be at the mercy of his habits, as he will find out to his dismay when a fresh or unusual situation does occur. And to be at the mercy of one's habits is to be out of control, that is to say, in bondage.[31]

In other words, the habitual responses that man learns to use in coping with life's difficulties begin to constitute a more subtle and potentially more hazardous difficulty. Habitual learned responses continue long after the conditions that engendered them have disappeared. And they in turn engender new habits that must themselves be unlearned if one is to operate as a "free" and responsible agent. Potter unself-consciously blends the worlds of mid-twentieth-century psychology and ancient Indian metaphysics when stating that "this round of habits breeding habits is a part of what is called in Sanskrit *saṃsāra,* the wheel of rebirth, which is governed by *karma,* the habits themselves."[32] It is not necessary, according to this view, to believe in transmigration, or an immortal soul, to see the advantage of attaining yogic self-control. Karmic bondage is, for Potter, a problem inherent in all human experience and is not to be thought of as a peculiarly Indian invention.

These discussions of habits and control strictly follow behaviorist principles, and although Potter does not specifically use Skinnerian technical nomenclature, theories of operant conditioning are never far beneath the surface of his discussion. Potter portrays karmic habits as conditioned responses: learned skills and traits that are acquired under socially determined problem-situations. To learn how to "succeed in the affairs of the world,"[33] specific patterns of behavior are selected and strengthened according to their success in reacting to various life-challenges. In Skinnerian terminology, these habitual responses are "reinforced" by their ability to bring about certain desired results ("positive reinforcement") and to avoid undesirable conditions ("negative reinforcement"). Potter's description of the karmic mechanism precisely follows the behaviorist's model: pairing particular behavioral responses with specific stimulus conditions.

In this behaviorist's view of Indian culture, the world of *saṃsāra* is a world of controlled, acquired responses elicited by the types of stimuli he calls "challenges." An individual's karmic background can be equated with what a psychologist might call his "reinforcement history." He carries with him the habits and skills conditioned by his previous learning experiences and applies them to new situations that he judges, through the identification of similar "discriminative stimuli," to be similar to previous "stimulus-situations."[34] This is what Potter means by "habitual"—"habits" are acquired behavioral responses about which the individual has little choice. He has been "taught," "trained," and "caused" to respond in a particular fashion. His determining past—his reinforcement history—is his karmic burden.

Potter understands bondage as the inability to overcome habitual responses as new situations arise. In this circumstance the determined individual can respond only with behavior learned in previous experience. Although such learned responses are often effective, the individual is not acting in a totally free manner "because habits of mind and action are set up within him through his apparent success which, as it develops, he is unable fully to control."[35] Every uncontrolled conditioned response, "good, bad, or indifferent—binds a man as surely as another."[36] To be free, one must overcome the compulsion of one's own reinforcement history, and assert control over one's actions, at least to the extent that they are not programmed by previously determined habits (*karma*). Thus, Potter sees the essence of liberation as the traditional Indian attitude of "renunciation"—the ability to gain mastery over any situation combined with a lack of concern for personal gain or attachment to the result of one's efforts. The free individual, the *yogi,* should be able to respond to stimuli without concomitant reinforcement:

> One does not by renouncing, give up his ability to exert a certain capacity that he knows he has, but rather chooses not to be attached to whatever gain or loss might result from exerting that capacity. Attachment breeds bondage, habits which control the self and limit its freedom.[37]

Only when an individual is in control of all situations, and is not himself controlled by concern for reward or punishment, is he truly free from karmic determination. For Potter, renunciatory action is action without reinforcement—freedom that cannot be reached without overcoming one's entire history of conditioning. The free man is released from behavioral causation and still retains the capacity to succeed in all his encounters with the external world.

Potter's understanding of human behavior as a primarily causal phenomenon is most clearly expressed in his discussion of two doubts that may accompany the attempt to attain *mokṣa*. These doubts, termed "skepticism" and "fatalism," represented by the philosophical schools known as Cārvāka and Ājīvika, are, according to Potter, the two extreme positions in the spectrum of Indian views on freedom:

> The parallel fears are, on the one hand, the fear that nothing one can do can bring about a hoped-for result, and, on the other, the fear that nothing one can do can alter what is bound to occur.[38]

Skepticism is the doubt that events are regularly connected; consequently, the skeptic fears that sufficient conditions might not exist for control of the external world and personal fate. It is a doubt about causality, about whether it can produce the desired state of freedom. This is the doubt that corresponds to what Potter calls the state of "freedom to," the fear that nothing anyone could do would suffice to free him to effectively control his world. On the other hand, the fatalist doubts that man is free from a fate predetermined by impersonal forces independent of his own control. This fear corresponds to what Potter calls "freedom from." It represents a fear that necessary and sufficient conditions for events that lead to freedom are not in one's control. Or, in behavioral terms, fatalism is the fear that man is irrevocably determined by his reinforcement history, that he is "programmed" by his pattern of past conditioning.

Potter's description of these fears indicates his virtually exclusive interest in the causal nature of human existence. To become free, man must learn to control external causal chains and, at the same time, to extricate himself from causal chains of reinforced habits that determine his own behavior. It is Potter's contention that the Indian philosophers were concerned with attaining a liberated state in which the dual freedoms "to cause" and "to be uncaused" can be exercised. Individuals who have not attained this specialized type of freedom are, according to Potter, unable to act spontaneously or without the tyranny of external control.

Potter believes that the subject of causality is the central issue in

Indian philosophy. Consequently, the centerpiece of his book is an elaborate, schematic framework arranging all *darśanas* by their views on causal questions.[39] And in this framework, Nāgārjuna's school is described as a "leap" philosophy—a view that (like skepticism) denies that there are specific causal chains leading to complete freedom, but that (unlike skepticism) does not deny that freedom is possible. Most Indian schools, Potter contends, assert "progress" philosophies—they "specify causal relations between complete freedom (freedom-from-cum-freedom-to) and its necessary and sufficient conditions, which allow men to enter a causal series leading to complete freedom."[40] Nāgārjuna, unlike other Indian thinkers, refuses to characterize liberation as the end product of steady progress along a path of causes and effects. Instead, his philosophy is a "noncausal theory" (*ajātivāda*), which teaches that although *mokṣa* is attainable, the means for attaining it cannot be characterized as a causal chain.

For Potter, as for so many scholars before him, Nāgārjuna's status is problematic. If *karma* is the basic determinant of human life and if *mokṣa* is its ultimate goal—the goal toward which men must strive if they want to achieve full self-mastery—then Nāgārjuna's refusal to admit the potency of efficient causality puts him (in Potter's eyes) in an apparently paradoxical position:

> Is Nāgārjuna a skeptic? No, since he allows that causality has a limited play—that is what the dialectic itself shows . . . Nāgārjuna is not anti-rational; in fact, he elevates reason to the position of the prime means of attaining freedom. Unlike skepticism, his is a philosophy of hope: we *can* achieve freedom by our own efforts, through remorseless application of the dialectic. . . . Yet freedom is release from the conceptual, for Nāgārjuna as for all Buddhists. This seems to be an insoluble paradox.[41]

For Potter, Nāgārjuna's limited interest in causality (like all other concepts, he admits its validity only on the level of *saṃvṛti-satya* but not *paramārtha-satya*) opposes his own basic presuppositions. Potter's approach is to construct explanations, to organize the rules of the game. But as he says himself, in Nāgārjuna's view

> no explanation is necessary for something being experienced. In fact, no explanation is necessary for anything, since there is nothing to explain. . . . That is the clue to Mādhyamika—it doesn't try to explain. The challenge of explaining the world has been abandoned as

> not worth attempting to meet. Nāgārjuna has no theory of relations or of error—he has no theories at all.[42]

Unlike the early Orientalists, Potter refuses to dismiss Nāgārjuna as a mystic or a nihilist, although he indicates that this is exactly how most Hindu thinkers conceived of the Mādhyamika philosopher.[43] And like Stcherbatsky and Murti he is tempted to characterize Nāgārjuna as a "transcendentalist."[44] But his interests are not in the distinction between appearance and reality. He is interested only in causal relations and in causal explanations. As a result of this exclusive emphasis he finds very little to say about Nāgārjuna and so much to say about behavioral paths to freedom. He devotes long sections of his book to the Sāṃkhya, the Nyāya, the Vedānta, and the Buddhist Logicians. But Nāgārjuna is mentioned only briefly and cautiously when he is mentioned at all (there are only five pages devoted specifically to Mādhyamika in the entire book). Potter is indecisive in his judgments about Nāgārjuna's philosophy:

> The Mādhyamika's position is frequently pictured by its opponents as one of skepticism, and it is difficult to know whether this is an accurate picture or not. . . . Mādhyamika is best construed, I think, as a leap theory.[45]

Frustrated, he warns his reader that Nāgārjuna does not seem to "believe in the power of the human mind to unravel the mysteries of the universe."[46] Potter essentially defines philosophy as the human attempt to unravel and demystify—that is, for him, the whole point of the game. Disappointingly, Nāgārjuna does not seem to want to play at all, though he does want to claim its prize: freedom from karmic bondage.

Potter's behaviorist leanings and his insistence on causality as the central subject of Indian thought reveal a positivist bias at least as thorough as Richard Robinson's (leavened with a certain amount of existentialist concern for the issue of freedom). Like Robinson, he displays a fascination with formulas, charts, and arguments, and like Robinson, he reconstructs the Indian tradition, while unwittingly repressing large portions of it, to fit his own presuppositions. For both of these scholars, as for other members of their generation, the ideals of scientific objectivity and analytic expertise are of determining importance. From logical atomism, Vienna Circle positivism, logical empiricism, and other types of analytic philosophy, these Indianists borrowed an aesthetic appreciation for scientism and a reverence for the ideal of neutral "logical," "mathematical," and "scientific" systems of expression and interpretation. The Russellian idea that logic is the "essence of philosophy" exerts stylistic control over the writings of both philosophers and nonphilosophers in any number of scholarly fields. The analytic readers of Indian philosophy, "properly armed with the technical apparatus of logic,"[47] as B. K. Matilal unself-consciously puts it, intend to substitute a purer, more scientific language for the poetry of idealism and even for the rich confusion of the Sanskrit originals. The problem is really a problem of language. This would, at least, be the supposition of the next generation of interpreters.

4

Buddhism after Wittgenstein

A New Game

Like their philosophical brethren, analytic Asianists such as Richard Robinson and Karl Potter were concerned with replacing the metaphysical language of their predecessors with more precise and scientific forms of expression. But these analytic interpreters of Nāgārjuna's verses, although determined to translate classical Sanskrit into logical symbols and mystical soteriology into behavioral science, were unable to comfortably claim Nāgārjuna as one of their own. Although he could be portrayed as a dialectician, Nāgārjuna's apparent doubts about the validity of rationality and the reliability of linguistic representation clearly made him something other than an early Indian equivalent of a Russell, an Ayer, a Carnap, or the Wittgenstein of the *Tractatus*. Instead of presenting a theory of knowledge, Nāgārjuna appeared to deny that either theories or knowledge were of ultimate value. He did not permit a firm distinction between facts and the language in which they might be expressed. And he seemed to hold a woolly relativistic position on truth in general. This kind of view was, at the very least, confusing to the analytic philosophers who tried to make sense of Nāgārjuna's writings.

Whether they sided with the thesis or not, all interpreters agreed that Nāgārjuna seemed to doubt that language could express or describe reality with any degree of exactitude. As Robinson worriedly observed, for Nāgārjuna, all language had only a limited, functional value. Even the most important terms in the Mādhyamika canon, such as *śūnyatā* and *paramārtha-satya,* were not meant to "name" existent entities or states:

> Emptiness is not a term outside the expressional system, but is simply the key term within it. Those who would hypostatize emptiness are confusing the symbol system with the fact system. No metaphysical fact whatever can be established from the facts of language.[1]

It was this position that particularly troubled many members of Robinson's generation. That reality is inexpressible (labeled by Matilal as a "dominant idea of all mystical philosophers"[2]) was a difficult view for most mid-twentieth-century scholars to sympathize with. For those enamored with the scientific "feel" of early analytic philosophy, beliefs, or propositions, must necessarily be either true or false. And precise, technical languages were imagined that could call every particular in the world by its particular name. The world was, for these thinkers, a collection of facts that could be empirically verified and indubitably grounded in a system of justified, true belief. But Nāgārjuna appeared to insist that there were no facts. It seemed that he would have had little regard for the project of attempting to express truth in formalized languages. Consequently, he was still labeled a mystic by the very people for whom mysticism was a euphemism.

But as with other intellectual fashions, the dominance of positivism was not permanent. The urge for representing, defining, denoting, quantifying, and picturing began to wane, and the emphasis shifted to more contextualist methodologies. As soon as terms such as "language game," "family resemblance," "private language," "form of life," and "ordinary language" started to filter into the conversations of students of Indian philosophy, Nāgārjuna's name was immediately, and repeatedly, linked with Wittgenstein's. Unlike the eighteenth- and nineteenth-century interpreters who had dismissed Nāgārjuna as a nihilist, the idealist metaphysicians who had seen him as describing the split between appearance and reality, or the logical-empiricist interpreters who had criticized him for insufficient logical rigor, this new generation of scholars began to read his *Mādhyamikakārikā* as if it were an explanatory gloss on the *Philosophical Investigations*. In many cases, Wittgenstein's remarks also began to be interpreted as if he might have been a Mādhyamika Buddhist:

> There is a close analogy between Nāgārjuna's rejection of self-conditioning entities and Wittgenstein's denial that any object can

> serve as its own criterion of significance in the private language argument. We can now extend our understanding of this common form by noting the similarity of Nāgārjuna's positive treatment of *pratītya-samutpāda* and Wittgenstein's notion of a "language game." . . . For Wittgenstein, a word or a perception of something has significance when logically connected into the criterial network of the language game. The private language argument explores the possibility that there might be elements identified independently of and prior to this network, like Nāgārjuna's *svabhāvas*. But this possibility is rejected. The rejection of atomic elements in the language system means that the elements must support each other mutually. This is exactly the sort of conceptual connection that Nāgārjuna calls interdependent arising.[3]

The "mutual interdependence" of all reality, the nonreality of isolated particulars ("atomic facts"), the unreliability of any and all types of linguistic constructs for precisely representing the world—these most troubling and slippery of Nāgārjuna's positions—suddenly begin to make sense (and more importantly, sound familiar) to the younger scholars who had read as much of Wittgenstein, Quine, Davidson, Sellars, and Kuhn as they had of Russell and Ayer. And instead of titles such as "Some Logical Aspects of Nāgārjuna's Philosophy" (Robinson), *Epistemology, Logic, and Grammar in Indian Philosophical Analysis* (Matilal), and "The Nature and Function of Nāgārjuna's Arguments" (R. H. Jones), studies now being published had titles such as *Wittgenstein and Buddhism* (Chris Gudmensen), "Nāgārjuna and Wittgenstein on Error" (Nathan Katz), and "Philosophical Nonegocentrism in Wittgenstein and Candrakīrti in their Treatment of the Private Language Problem" (Robert Thurman).

The reign of the analytic aesthetic had meant a relative dry spell for the old Jonesian methodology of comparative interpretation, particularly when it came to Nāgārjunian studies. Scholars such as Robinson and Potter diagrammed, charted, and symbolized their way through Mādhyamika texts, but found little with which to forge an identification between Nāgārjuna and contemporary analytic thinkers. But along with an enthusiasm for Wittgenstein's later writings, the comparative tendencies that had been prominent in the idealist decades once again found fertile soil in the new postanalytic period. This trend, like the idealist sensibility that had focused heavily on Advaita Vedānta, found Mādhyamika of central interest. Nāgārjuna was now suddenly, because of the comparisons with Wittgenstein, the most written-about Indian philosopher. And this was a comparison that went both ways. Wittgenstein's aphoristic style was being interpreted through Nāgārjuna's as much as Nāgārjuna's was being read through Wittgenstein's. As Murti had used the work of European transcendentalists to create an interpretation of Nāgārjuna that was intelligible to an earlier Western audience, these post-Wittgensteinian Asianists discovered that *The Investigations* allowed for a reading of Mādhyamika that felt up-to-date. Even more astonishingly, they felt that they could begin to make sense of some of Wittgenstein's difficult and highly controversial writings through the grid of Nāgārjuna's now familiar text. Many of them were even more comfortable with the ancient text, though in Sanskrit, than with the discomfortingly new text, available in side-by-side German and English versions.

Wittgenstein's often-mentioned private language discussions, Quine's notion of an interwoven "fabric of belief,"[4] and Davidson's thesis that truth was located in natural sentences rather than in isolatable propositions[5] are only the most prominent philosophical examples of a reaction against the analytic empiricism that had so dominated the first half of the century. And it was this new outlook, which stressed the linguistic contextuality of meaning and truth, rather than empirical verification, that made Nāgārjunian doubts about *svabhāvic* essentialism seem suddenly so contemporary. The perspectivism of Nāgārjuna's two truths, the antimetaphysical implications of his insistence on *śūnyatā,* and the stress on *pratītyasamutpāda* (a complex interdependence) as the basis of meaning were no longer seen as mystical or paradoxical. Wittgenstein's celebrated dictum that the "limits of language are the limits of my world" resonated for those familiar with Nāgārjuna's claim that "the limits of *saṃsāra* are the limits of *nirvāṇa.*" And Nāgārjuna's strange assertions about the ontological "emptiness" of the objects of language and conceptual thought sounded less strange to those who were now thinking of "meaning as use" rather than meaning as representation. The idea that our problems (and consequently, their solutions) could be found in our own beliefs about language's power to explain, describe, and name was everywhere. It was discussed in classes and professional conferences on contemporary Anglo-American philosophy and in papers and lectures on traditional Buddhism. Overlap between the two fields, at least in the work of professional Buddhologists, became increasingly commonplace:

> The parallel between the Buddhist notion of "convenient designation" and Wittgenstein's "everyday language" is clear. Both are saying that

> because a word may be *used,* we should not get carried away with philosophies about essences and the like.[6]

From this new vantage point, other parallels suddenly became clear as well. References to the "Russellian/Abhidharmist primacy of privacy"[7] and the "logical atomism" that "has as its counterpart in India the Nyāya school,"[8] were part of a new philosophical change. Both Wittgenstein and Nāgārjuna were portrayed as rebels against the tyranny of realist predecessors who were so philosophically barbaric as to have spoken of definition by ostension and to have assumed the existence of private sensations. Like Nāgārjuna, Wittgenstein was seen to offer liberation from this older, essentialist's view of the world:

> A picture held us captive.[9]

> What is your aim in philosophy?—To show the fly the way out of the fly-bottle.[10]

> Philosophy is a battle against the bewitchment of our intelligence by means of language.[11]

> Philosophy, as we use the word, is a fight against the fascination which forms of expression exert on us.[12]

And like Wittgenstein's disenchantment with logical positivism, Nāgārjuna was understood as establishing a philosophical critique of the Abhidharma Buddhist's attachment to elemental *dharmas* and the Naiyāyika empiricists' assertions of "is" and "is not" (*asti* and *nasti*):

> The error of Nyāya is to attribute to each *svabhāva*—here the antithesis of "dependently co-originated"—and to speak of them abstractly as reals.[13]

Wittgenstein and Nāgārjuna, increasingly being spoken of in one breath, as if they had collaborated on their writings, and as if they had the same philosophical adversaries and the same goals, were the focus of a new school of comparativists. Nāgārjuna, who had, a few years before, been identified so closely with the transcendentalism of Kant, Hegel, and Bradley, was now viewed as the leading ancient, non-Western proponent of the ordinary-language philosophy that was displacing verificationist empiricism in departments of philosophy. In the hands of one of its most influential proponents, Frederick Streng, a summary of the new Wittgensteinian Mādhyamika made it sound as if it were a philosophy that belonged as much at twentieth-century Cambridge as in second-century South India:

> Nāgārjuna's use of words for articulating Ultimate Truth would find champions in contemporary philosophers of the language analysis school. . . . Words and expressions-patterns are simply practical tools of human life, which *in themselves* do not carry intrinsic meaning and do not necessarily have meaning by referring to something outside the language system. Wittgenstein suggests that language is like a game, and the meaning of a word or phrase depends on the "rules" which one learns "to play this game." . . . Throughout the *Philosophical Investigations* Wittgenstein argues that metaphysical inferences are simply fabrications based on a misconceived notion about how meaning is available. The proper role of philosophy is to clarify the use of words as they are used in specific contexts rather than build "castles in the air." . . . In a manner similar to the contemporary language analyst, Nāgārjuna denies that all words gain their meaning by referring to something outside of the language system; he maintains that the relationship between words in a statement (e.g., subject and predicate; the person acting, the action, and the object acted upon) are only of practical value and not indicative of ontological status.[14]

The new interpreters were not equally emphatic in identifying the Wittgensteinian and Nāgārjunian programs. Streng, for example, is careful to differentiate between their audiences: Wittgenstein is concerned only with the mistakes of professional philosophers, whereas Nāgārjuna is concerned with the salvation of all human beings.[15] Gudmunsen, on the other hand, is determined to deny even this slight dissimilarity in approach and asserts, against Streng, that "there is not nearly as much difference between the roles of Wittgenstein and Nāgārjuna as one might imagine."[16] But despite fraternal disputes such as these, the new "Wittgensteinian" interpretation of Nāgārjuna's text quickly began to supplant older idealist and analytic readings. It was an interpretation built on the various forms of philosophical holism, relativism, and contextualism that were replacing the obsessions with scientific certainty and logical structure. The new obsession was with interdependence and in socially grounded criteria of meaning—an interweaving of *pratītyasamutpāda* and "meaning as use."

***Pratītyasamutpāda* and the Philosophy of Language**

The modern, multivolumed "histories" and "outlines" of Indian philosophy, by respected authorities such as Warder, Frauwallner, Hiriyanna, Bernard, and Raju, were written under the collaborative influences of

European twentieth-century logical empiricism and the Indian traditional model of competitive metaphysical sects. Like the earlier idealism-tinged monument of encyclopedic scholarship, Dasgupta's *A History of Indian Philosophy,* their general tendency was to describe individual speculative schools advancing opposing theories about the nature of ultimate reality. Nāgārjuna's place in this assemblage of systematic theories was, as in the work of the positivist Buddhologists, considered to be ambiguous. During the period in which the aesthetics of logical and conceptual analysis controlled the range of possible approaches, Mādhyamika was recognized both as an idiosyncratic logical critique of the various metaphysical and epistemological positions existing at the time, and as an irrational, soteriological mysticism. But the changes in this ambivalent view after the new post-Wittgensteinian influences began to be felt by Indianists were very dramatic. As the sharply demarcated distinctions of logical empiricism began to give way to fuzzy boundaries that allowed more room for apparent paradox, concern for semantic representation was discarded in favor of "family resemblance" and the desire for definition gave way to talk about interpretive discourse and "forms of life." The Western understanding of the Mādhyamika school, as well as the general outlines of the history of Indian thought, took on a different shape due to the Wittgensteinian idea that the meaning of an utterance was a matter of its use within an appropriate discursive context.

Śūnyatā and *nirvāṇa* had been the most frequently used Sanskrit terms in the work of Stcherbatsky and other early interpreters. As terms for describing the Mādhyamika absolute, they were essential to the transcendentalist interpretation. Similarly, *svabhāva* was central for an "analytic" interpreter such as Robinson. The logically inconsistent essentialist's insistence on "own-being" was, in Robinson's view, the chief target of Nāgārjuna's dialectical strategy. The new predominance of *pratītyasamutpāda* (most often translated as "dependent coorigination") in the writings of the post-Wittgensteinian Nāgārjuna scholars marked more than just a shift in terminology. It was an important change in the philosophy of interpretation. For the first time Mādhyamika was being read as a metaphilosophical critique of the language of philosophy. And consequently, the philosophical presuppositions of the interpreters were considered to be as important in understanding Nāgārjuna's text as the philosophical position of the author. Western scholars of the Mādhyamika canon were now concerned with self-consciously avoiding the interpretive extremes of both transcendental absolutism and radical nihilism.[17] From this standpoint, all earlier readings were viewed as having a kind of isogetical essentialism: insistence on a univocal and static notion of truth in the original text. Consider Streng's clear statement of this new interpretive policy:

> The "emptiness" which denies any absolute, self-sufficient being also establishes existence (i.e., existence empty of any self-existent reality) through dependent co-origination; emptiness is neither an absolute monism nor nihilism.[18]

For readers such as Streng, who are self-avowedly disposed toward Wittgensteinian reformulations of philosophical issues, the idea that all terms are totally interrelated and consequently semantically underdetermined by any particular discursive context is neither problematic or "paradoxical." Misinterpretations, they feel, resulted because previous readers of the *Mādhyamikakārikā* were not fully aware of the contextual nature of philosophical language and consequently did not appreciate the importance of the interrelatedness (*pratītyasamutpāda*) and relativity of linguistic expression and of all philosophical issues. And, more importantly, scholars such as Streng, Thurman, and Katz argue that earlier interpreters had erred in refusing to take Nāgārjuna at his word about the status of his own teachings. The language of his text, like all other conventional human discourse, derives its meaning from its place within a system of functional interdependence. Nāgārjuna, they believe, "seems to be telling us that our words and concepts neither adequately explain nor describe reality."[19] And this, they feel, echoes Wittgenstein's dictum that "a word is not necessarily the name of a thing, and that philosophic problems arise when one abstracts a word from its context—that is, from the everyday workings of language—and tries to understand that word apart from its functional matrix."[20] His own words cannot be treated as somehow immune to the interdependent nature of conceptual thought and conventional language; therefore, they should not be treated as if they were meant to explicitly "name," "refer to," or "describe" anything on the level of absolute truth.

From this point of view, the characterization of Mādhyamika as nihilism, so common throughout the school's history, is the result of the error of taking the negativistic Mādhyamika terminology (*śūnya*) as well as the Nāgārjunian dialectical strategy out of its functional context. Nāgārjuna, they contend, based his methodology on a well-documented, traditional Buddhist position on the mere conventionality of language and thus could not have intended to directly name or refer to reality as "empty" or "void." It is void only to the extent that it must be empty of static "own-being" (*svabhāva*). And more importantly, it is *śūnyatā,* it is "emptiness," that is necessary for reality to exist at all. Chapter XXIV of the *Mādhyamikakārikā,* once so important to the idealist interpreters at the beginning of the century, becomes central again to the Wittgensteinian scholars. It is in this chapter that Nāgārjuna states that "emptiness" (*śūnyatā*) is, in fact, that which allows what we perceive as existence to exist. Without *śūnyatā,* and without participation in *pratītyasamutpāda,* there could be no reality, void or otherwise.

> Any factor of experience which does not participate in relational origination cannot exist. Therefore, any factor of experience not in the nature of *śūnya* cannot exist. (Verse 19)

If, as these readers contend, a central tenet of the Mādhyamika position is that language cannot define or precisely encompass ultimate reality, then the term *śūnyatā* is, like any other word, no more than a conventional formulation for describing the unreality of the philosophical notion of own-being. The closest that Nāgārjuna comes to a "description" of reality is the term, *pratītyasamutpāda,* which implies only relativity and complex interdependence. In this view, "emptiness" does not name the true condition of reality or imply that all existence is unreal or nonexistent. Katz writes that "both Wittgenstein and Nāgārjuna are explicit in stating that in their negations they are not negating anything real—which is consistent with our characterization of them as both being positionless."[21] According to this view, Nāgārjuna would agree with Wittgenstein's statement that "what we are destroying is nothing but houses of cards and we are clearing up the ground of language on which they stand."[22] And, Katz contends, the common misunderstanding of Mādhyamika as nihilism happens

> precisely because their negations are taken as ontological rather than grammatical, leading to the false idea that Nāgārjuna negates reals . . . Nāgārjuna's negation is a corrective, an inquiry into the grammar of talking about the world, while the nihilist's negation is, according to the Mādhyamika, just another view (*dṛṣṭi*).[23]

Just as this new understanding of Nāgārjuna as a philosopher of language was used to combat the interpretation of nihilism, the alternative misunderstanding of Nāgārjuna as an absolutist, can, according to this view, be similarly dismissed. Building on Robinson's understanding of Nāgārjuna as a critic of philosophical absolutisms, Streng, Katz, and the others contend that *Śūnyatā* certainly does not refer to some existent entity or condition that could serve as an absolute reality for Nāgārjuna's system. The monistic and absolutistic interpretations are viewed by these scholars as no less mistaken than the nihilistic readings. From the first verses of his text, they argue, Nāgārjuna explicitly denies any ontological first principles or metaphysical substrata. And, once again, a traditional Buddhist position is invoked: the refusal to discuss metaphysically oriented issues. According to the Pāli and Abhidharmika literature, the Buddha is said to have declared metaphysical and cosmological questions to be unanswerable. The parable of the man who, wounded by a poison arrow, did not wish to be treated for his injury, but wanted only to know the origin, composition, maker, archer, and so on of the weapon, was canonically indicative of Buddhism's preference for soteriology over ontology. Metaphysical questions must be ignored, it was taught, in favor of religious salvation and the kinds of "truth" necessary for that salvation. And this was not just an indication of the ineffability or inexpressibility of ultimate reality, which had been the transcendentalist's reading of the Buddha's "silence" on cosmological and metaphysical questions.[24] For Nāgārjuna, scholars now argued, metaphysics is rendered irrelevant by a method of analysis that attempts to reduce all presupposed philosophical premises to absurdity by revealing essentialist habits of linguistic expression.

The avoidance of absolutistic ideas is thus a methodological constant in the Mādhyamika system. Although Nāgārjuna does speak of the "two truths," this bifurcation is not meant to function, it is maintained, as a disjunction between two levels of metaphysical or epistemic reality. It is a distinction between two ways of talking: two "language games." There is no effort to posit an absolute reality beyond that of the conventional; there is no theory of an ineffable, supernal realm of absolute existence that stands outside the reaches of conceptual thought and conventional reality. In this regard, it is significant, as scholars who portray Mādhyamika as a philosophy of language begin to point out, that Nāgārjuna speaks only of the "two truths" (*paramārtha-satya* and *saṃvṛti-satya*) and not the "two realities." Any understanding of *śūnyatā* as a metaphysical absolute is contrary to Nāgārjuna's intentions and is a result of confusion between levels of the linguistic and nonlinguistic orders.

> By clearly understanding that there is no absolute essence to which "emptiness" (or "*nirvāṇa*" and "perfect wisdom") refers, we recognize that when emptiness is described as inexpressible, inconceivable, and devoid of designation, it does not imply that there is such a thing having these as characteristics.[25]

Because Nāgārjuna views his own language as merely conventional, and as empty itself, the new interpreters argue that absolutism and nihilism can both be seen as symmetrically misleading interpretations. Avoidance of the two ideological extremes thus becomes a methodological goal in itself.

As part of his examination of Tibetan Mādhyamika sources, for example, Robert Thurman translates the writings of the fifteenth-century Prāsaṅgika Mādhyamika philosopher, Tsong Khapa, on precisely this issue. From his major text, in a chapter entitled "The Essential Mādhyamika Message," Tsong Khapa declares that the philosophical views that must be discarded to achieve enlightenment are the views of "eternalism" (absolutism) and "nihilism."

> Those who do not believe in the relativity which is the dependent origination of persons and things, but believe in the truth [-status] of those two . . . fall into the abyss of the eternalistic or nihilistic views.[26]

He further states that it is belief in the ultimate reality, the absolute truth, of conventional entities and philosophical ideas that leads to the adoption of either absolutistic or nihilistic ideologies.

> Eternalism is eliminated by the fact of the non-establishment in reality (of persons and things), and nihilism is eliminated by the fact of the functional capacity of each thing.[27]

According to this version of the Mādhyamika teachings, absolutistic ideas are negated by Nāgārjuna's discussion of the "emptiness" of all individual entities and concepts. Nihilism is similarly combated by the lack of an ontological denial of common-sense reality; conventional reality as established by the relativity of voidness. The two extremist views, according to this interpretation, are the poles through which the Mādhyamika seeks a "middle way." Nāgārjuna's aim is not to deny conventional knowledge, common-sense assertions, or the conventional validity of "ordinary language," but to deny chimerical notions of intrinsic reality and essential being that can become attached, through philosophical language, to common-sense assertions.

Attention to the entire text of the *Mādhyamikakārikā* serves to establish the importance of *pratītyasamutpāda* for this new school of interpreters. In the first 21 chapters of the text, as Robinson had already so carefully demonstrated, Nāgārjuna had attempted to demonstrate that our understanding of entities or concepts that might be thought to possess intrinsic identity are actually composed of two or more contingently related ideas. If all possible instances of *svabhāva* can be converted into interrelated dyadic or triadic relations, they can be shown to be "void," to the degree that they cannot be said to truly participate in simple, independent, ultimate reality. Consequently, visual perception has two components: the "seeing agent" and the "seeing activity" (Chapter III). Volitional action is described as composed of "the doer" and "the deed" (Chapter VIII), burning as an inseparable relation of fire and wood (Chapter X), and the normal, human experience in *saṃsāra* is shown to be a relative state between unreal extremes of "bondage" and "release" (Chapter XVI). In this manner, our concepts of conventional phenomena are understood as mutually interdependent.

A similar, but slightly more complex, formulation that Nāgārjuna introduces in Chapter II, the analysis of time as a continuum divided into three related components, is a further variation on this component analysis argument used for simple pairs of relata. Any concept of time, change, or motion (Chapters II, VII, XI, and XIX) can be broken down into the three moments of prior, current, and posterior occurrence. This argument (analogous in form to Zeno's hypotheticals about infinite divisibility) is then used to demonstrate infinite regress: if any moment has three necessary submoments of past, present, and future, then each of these submoments is also composed of further submoments—and so on, ad infinitum. The ordinary concept of time as a single, absolute continuum is thus seen as being inherently more complex and less conceptually reliable than would ordinarily appear.

These sorts of dialectical strategies were already being identified by several analytically minded scholars without much approval or enthusiasm (as in the work of Robinson and Matilal). But Nāgārjuna's attempt to inspect all entities, qualities, and concepts and to find them to be "void with respect to absolute reality" was reinterpreted by scholars who were also readers of Wittgenstein. Now, this component analysis of conventional phenomena, combined with a new appreciation for the significance of the relativity of ideology, language, and perception allowed a far more positive response. Nāgārjuna now seemed to argue that, like utterances in a language game, nothing can be shown to exist except as a participant in a complex network of interrelations. Cause and effect, an entity and its properties, an agent and its actions—all are shown to be empty, or only conventionally true. No entity or concept is "self-existent" or possesses "essential being." It is only because things do participate in *pratītyasamutpāda,* because their only nature is *śūnya,* and because they do *not* have a nature of *svabhāva,* that they can be said to exist at all. Without interrelatedness, the new interpretation states, Nāgārjuna would be unable to differentiate between the conventional qualities and objects of everyday existence and the imagined philosophical concepts (like *svabhāva*) of the essentialist philosophers. And this view, that philosophical mistakes are the result of a lack of self-consciousness about language use and a lack of attention to causal and conceptual relativity, is equated with Wittgenstein's celebrated views on the linguistic nature of philosophical "mistakes":

> The opponent, as the interlocutor in the PI passage, is a philosophical absolutist, a substantivist, who is "bewitched by language" into perceiving things to be absolutely true, "really real" before him, and the Wittgensteinian and Mādhyamika nonegocentrist critical analyses intend to force him to look deeper into things and processes by examining his account of them to actually try to find the essence to correspond to the name, the "metaphysical entity," the "simple," the "indivisible." The absolutist's failure to find any such analysis-resistant essence is the first step on the road to liberation of his intelligence from the spell of language. The century- and culture-spanning similarity of therapeutic technique is startling.[28]

The theme of freeing language from the absolutist's insistence on simple, indivisible concepts and entities is, along with a corresponding insistence on relativity and interdependence, the battle cry of the Wittgensteinian Nāgārjunians.

The understanding of Nāgārjuna as a critic of philosophical language serves further to begin to demystify the seemingly paradoxical twin ideas of "the two truths" and the infamous assertion of the identity of *saṃsāra* and *nirvāṇa* in Chapter XXV, verses 19 and 20. In Chapter XIII, Nāgārjuna had stated that mental constructions are ultimately delusive and, in Chapter XVI, this idea is more fully expanded in a way that implies both identity and difference. Nāgārjuna argued that both *nirvāṇa* and *saṃsāra* are beyond conceptualization, impossible to grasp through mental constructions. In this way, both realms are described as empty from an absolute perspective (*paramārthasatya*) and, as described in later chapters, indistinguishable from each other. But from the conventional perspective of linguistically determined conceptualization (*saṃvṛti-satya*) the disparity between conventional and absolute truth is wide indeed. This controversial claim that things are simultaneously the same and different is less paradoxical, the new interpretation holds, when it is understood as a "grammatical rather than an ontological statement."[29] This is a matter of a distinction between two possible perspectives and two differing linguistic contexts. It is true that Nāgārjuna argued that things cannot conventionally be said to have absolute, intrinsic nature in themselves, but this is a caution about, as Wittgenstein had written, "the limits of language," not a speculative theory about essence or being. Nāgārjuna's description of mental constructions and language as empty is intended, the Wittgensteinians point out, to highlight the merely functional value of philosophical expression, the lack of a supposed correspondence between words and the meanings they are purported to have by such realists as the Naiyāyika denotation theorists.[30]

> To say that things have *svabhāva* is to say that one can coherently speak of them apart from their everyday language, that a word has a referent, which is to say that a word is more than a convenient designation.[31]

As a critique of philosophical positions and a critique of philosophical language, the Mādhyamika "nonposition" is not meant to establish a metaphysical distinction between the immanent and the transcendent, an epistemological distinction between *noumena* and *phenomena,* or a hierarchical division between realms of spiritual reality. According to a post-*Philosophical Investigations* interpreter such as Thurman, Nāgārjuna is thus seen as a distant precursor to Wittgenstein on the subject of the conventionality of language and the social construction of philosophical ideas. He writes that the Mādhyamika school's primary target, like Wittgenstein's, was

> the classic absolutist mistake of thinking that lack of an *absolute* basis is no basis at all, lack of an *absolute* process is no process at all, lack of an absolutistic, privately grounded language is no language at all, lack of a mathematically absolute, perfect logic is no logic at all, and so on.[32]

As is customary for this new comparative moment, Thurman's writings on Mādhyamika, like Streng's, Katz's, Gudmunsen's, and others, offer parallels with, and quotations from, Wittgenstein's later work. Thurman also compiles an extensive list of quotations from Wittgenstein on the subject of the conventionality of language, which includes many of the same quotations that repeatedly appear in the writings of all the Wittgensteinian interpreters:

> (About) the 'language of our perceptions', . . . this language, like any other, is founded on convention. (PI 355) . . . One objects: "So you are saying that human agreement decides what is true and what is false?"—It is what human beings *say* that is true and false; and they agree in the *language* they use. That is not agreement in opinions but in form of life (PI 241) . . . Here we strike rock bottom, that is, we have come down to conventions. (BBB, p.24) . . . When philosophers use a word—'knowledge', 'being', 'object', 'I', 'proposition', 'name'—and try to grasp the *essence* of the thing, one must always ask oneself: is the word ever actually used in this way in the language-game which is its original home. What we do is to bring words back from their metaphysical to their everyday use. (PI 116) . . . The meaning of a word is its use in the language. (PI 43) . . . When I talk about language (words, sentences, etc.) I must speak the language of everyday. Is this language somehow too coarse and material for what we want to say? *Then how is another one to be constructed?* (PI 120) . . . A main source of our failure is to understand is that we do not *command a clear view* of the use of our words. (PI 122) . . . Philosophy may in no way interfere with the actual use of language; it can in the end only describe it. For it cannot give it any foundation either. It leaves everything as it is. (PI 124) . . . *Essence* is expressed by grammar (PI 371). Grammar tells what kind of object anything is. (Theology as grammar) (PI 373). [33]

Thurman selects these quotations to constitute an interpretation, even a restatement, of Candrakīrti's, Tsong Khapa's, and Nāgārjuna's Mādhyamika. The new understanding, following the period of positivist enthusiasms for formal logic and scientific algorithms, is that Wittgenstein's emphasis on language could serve to explain Nāgārjuna's previously inexplicable teachings about "dependent co-origination" and the multivocality of truth. The two philosophers, speaking from widely different cultures and different ages, were both concerned, so Streng writes, "to clear away misunderstanding about the use of words."[34] Both appeared to be involved in Wittgenstein's "battle against the bewitchment of our intelligence by means of language."[35] Both appeared to insist that, as Nāgārjuna had put it, "a wrongly conceived *śūnyatā* can ruin a slow-witted person. It is like a badly seized snake or a wrongly executed incantation."[36] The enemy of these two philosophers seemed to be nothing other than philosophy.

The End of Philosophy?

The philosophical era I am calling post-Wittgensteinian (which can just as easily be termed post-Heideggerian, post-modern, or simply, post-Second World War) is characterized by the idea that it is more self-conscious than earlier periods and by its suspicion of any and all theoretical commitments. Among philosophers of this persuasion, philosophical speculation is itself often suspect and the task is to uncover or "de-construct" the presuppositional fallacies on which the age-old arguments and obsessions are based. Philosophy also no longer appears to hold the same allure as it did for Asianists influenced by methodologies of "conceptual analysis" and "analytic definition." Streng, for example, stresses that his study, *Emptiness: A Study in Religious Meaning,* is concerned with "religious truth," "religious knowledge," "religious meaning," and "religious awareness." The adjective, "religious," is placed consistently and carefully before every noun that might be construed as "philosophical" in nature. As Streng writes,

> "Emptiness" has been most often investigated as a philosophical term. There is certainly sufficient material in the texts of Nāgārjuna and those of his disciples with which to construct a systematic presentation of such perennial problems as the nature of cause and effect, reality, existence, and knowledge. We need only to recall the studies of T. Stcherbatsky, A.B. Keith, T.R.V. Murti, and E. Frauwallner, to suggest the work that has been done from a philosophical perspective. . . . All these men are specialists to whose considered judgments we must resort; nevertheless, my own concern with the material is somewhat different.[37]

Rather than the traditional philosophical perspective, or the predominantly historical treatment given to Mādhyamika materials by non-philosophers such as La Vallée Poussin, Dasgupta, and Lamotte, Streng's concern is with investigating the relation between religious awareness and forms of verbal expression. Unlike Robinson's logical aggressiveness or even Matilal's late-analytic forays into conceptual analysis, Streng claims that he intends to understand emptiness as part of a "religious language game," as an expression of a "religious form of life." Besides Wittgenstein (whose work is not cited nearly as frequently by Streng as by some of the other writers quoted here) names of phenomenologists and historians of religion such as Rudolf Otto and Mircea Eliade are invoked as methodological authorities, not the names of Anglo-American logicians and epistemologists.[38]

Analytically oriented writers had previously suggested, of course, that Nāgārjuna's position was neither nihilism, absolutism, nor any other kind of familiar philosophical ideology. We have seen that Robinson uncomfortably but open-mindedly asserted that *śūnyatā* was not a term that stood outside of the conventional system of discourse and that "hypostatizing" of any term within Nāgārjuna's system was inherently misguided:

> The term "absolute truth" is part of the descriptive order, not part of the factual order. Like all other expressions it is empty, but it has a peculiar relation within the system of designations. It symbolizes non-system, a surd within the system of constructs.[39]

Murti, for all of his own commitment to the transcendentalist tradition, had written that Nāgārjuna had not attempted to offer a speculative theory or an ontological position:

> Philosophy, for the Mādhyamika, is not an explanation of things through conceptual patterns, that is the way of dogmatic speculation (*dṛṣṭi*); but this does not give us the truth. The Dialectic is intended as an effective antidote for this dogmatic procedure of reason; it is the criticism of theories (*śunyata sarva-dṛṣṭīnām*). Criticism of theories is not another theory; *śūnyatā* of *dṛṣṭis* is not one more *dṛṣṭi*, but is *prajñā*—their reflective awareness.[40]

But just as Murti was sure that the Mādhyamika dialectic really sought to uncover the fact that "the Absolute (Śūnya) is the universal impersonal reality of the world,"[41] Robinson had been similarly convinced that there must be a "fact of the matter," or at least a consistent formal structure, behind the unsatisfyingly mystical paradox of "emptiness." Perhaps the most contemporary of the analytic interpreters, Bimal Matilal, also offers a discussion of the Mādhyamika system that indicates considerable tension between the pre-Wittgensteinian expectations of positivist scientism and post-Wittgensteinian wariness of explanatory theories. Although on the one hand offering a mildly disapproving, formal analysis of Nāgārjuna's dialectic as a fallacious logical system based on the principle of "predicative indeterminance," Matilal echoes Murti's statement that Mādhyamika was a critique of all views of reality, but was "not itself another view of reality":

> Emptiness should, under no condition, be construed as a view (*dṛṣṭi*) or a position. It has the therapeutic value of curing delusions originating from all sorts of views or positions . . . it would be foolish to construe "emptiness" as a position.[42]

These consistent assertions that Nāgārjuna did not either hold or want to hold a philosophical position, although repeated for years in the interpretive literature, were not stated with conviction or without misgivings until Nāgārjuna's "nonposition" began to be compared with Wittgenstein's. The famous Wittgensteinian cautions about philosophical "mistakes" and, most importantly, his portrayal of philosophy as a process of therapy offered a newly comfortable, interpretive context:

> The philosopher is the man who has to cure himself of many sicknesses of the understanding before he can arrive at the notions of the sound human understanding.[43]

From this "therapeutic" Wittgensteinian perspective, Nāgārjuna appears to have also constructed a system that looks more like a prescription for cure than a philosophy. The task of the Mādhyamika philosopher, in this view, is neither to escape to another realm of higher reality, nor to deny the reality of everyday existence. His goal is to eliminate delusive, conceptual habits of mind (to use the psychologistic construction) and to avoid holding absolutism, nihilism, or even *śūnyatā* as a philosophical position. *Nirvāṇa* is not to be thought of as the "real" reality, nor is *saṃsāra* explicitly denied as utter "unreality." Since *śūnyatā* is a self-referential term (suggesting that *śūnyatā,* too, is "empty") Nāgārjuna's *nirvāṇa* could not have been meant to be a separate realm of existence, nor *śūnyatā* as a "true" view of reality. *Nirvāṇa* is strictly a term for designating a state unencumbered by philosophical concepts and unmediated by the language of existence and nonexistence. And *śūnyatā* is a provisional, heuristic term for the nature of the universe as seen without such absolutistic mental constructions. It is reality without an "illusory veil" of conceptual formations that impose a conventional idea of reality. *Paramārtha-satya* is thus, in this description, a psychological state. According to Streng, absolute truth can be described, not as a philosophical view or collection of teachings, but as a special, aphilosophical state of awareness:

> (It is) living in full awareness of dependent co-origination rather than in a limited, 'tunneled' awareness about the conditions of existence. It is living without fear of the interdependent nature of things, and without desire for an unconditioned self-existent reality—which is just a fantasy, a mirage.[44]

Nāgārjuna's celebrated warnings about the perils of wrongly understanding *śūnyatā* ("Like a snake badly seized," Chapter 24, verse 11) or holding this nonposition as if it constituted a philosophical view in itself ("Those for whom *śūnyatā* is itself a theory, I call incurable." Chapter 13, verse 8)[45] thus become doubly important in the new Western climate of scepticism about philosophical theorizing. Nāgārjuna, like Wittgenstein, did not intend to substitute his theory for those of his opponents. His only intention, Katz writes, was to cure others of the philosophical illness:

> The philosophical, for Buddhism, is rooted in the psychological. The belief in self, which Buddhism also holds to be a grammatical mistake, is deeply rooted in our desire that there be a self. . . . Philosophic problems are a disease to be cured because life is a disease to be cured. . . . Philosophical problems do indeed arise when language goes on holiday.[46]

Streng and Katz both emphasize (more than do Thurman, Gudmunsen, or Waldo, for example) that Nāgārjuna and Wittgenstein appear to diverge when it comes to the scope of their proposed cures. As we have seen, although Wittgenstein may have been concerned only with the linguistic illnesses of a small group of people who deal with the problems of professionalized philosophy, Nāgārjuna was convinced, as a Buddhist, that the salvation of all living beings was at issue. For the Mādhyamika, the disease of essentialist language "is not only a problem for the philosopher. It is a disease with which we are all afflicted."[47] Thus, Nāgārjuna's effort, Streng writes, "is best understood in terms of a religious concern for release from suffering."[48] But this difference in therapeutic scope is not enough to dampen the enthusiasm for the idea that both Nāgārjuna and Wittgenstein were endeavoring to attain an awareness that could be defined as "the ultimate liberation from the 'fly-bottle' of perplexity."[49] Katz asserts that the

> parallel between the Buddhist notion of "convenient designation" and Wittgenstein's "everyday language" is clear. Both are saying that because a word may be *used,* we should not get carried away with philosophies about essences and the like.[50]

So, Katz is saying, even the apparent disparities between their positions may not be as great as might first appear. Wittgenstein may feel that "everyday language" is fine as it is, but Nāgārjuna similarly asserts that the "convenient designations" of conventional discourse are perfectly in order, as long as they do not begin to be used as *philosophy*. It is philosophical truth that seems to be the common enemy of both thinkers, and it is philosophical language that is the root of the problem.

For an entire generation of Buddhist and Indian scholars enamored with the *Philosophical Investigations* and *On Certainty*, the proposed parallels with Nāgārjuna's verses on the Middle Way offer a sense of urgency and historical immediacy that had been unavailable since the time of Schopenhauer's enthusiastic "discovery" and endorsement of the ancient Hindu precursors of nineteenth-century German idealism. *Pratītyasamutpāda*, *śūnyatā*, and *paramārtha-satya* could now be understood as antiphilosophical slogans in a modern battle against the "bewitchment of our intelligence" by the language of *svabhāva*. An analogue for the postempiricist disenchantment with "objective reference," "meaning analysis," and scientific models of truth had been found in Buddhist scriptures. Quinean notions about the "indeterminacy of translation," Kuhn's and Feyerabend's examples of the "theory-ladenness" of observation, and Davidson's reformulation of terms such as "truth" and "conceptual schemes" could now be discussed in relation to a second-century Buddhist thinker.

AFTERWORD

Holists, Hermeneuticists, and Holy Men

This text has not been about Indian philosophy, Mādhyamika Buddhism, or the history of Asian studies. Although it examines at length the Western scholarship on the teachings attributed to Nāgārjuna, these materials are meant to form a theoretically oriented case study. I have made no attempt to present a comprehensive survey of Nāgārjuna scholarship; I have neither included all works on the subject in my discussion nor written a proper history of the field. Major scholarly contributions, like those of Stanslaw Schayer, J. W. de Jong, and Kenneth Inada, and new, attention-getting works such as those of Robert Magliola (Part 3 of *Derrida On The Mend,* 1984) and David J. Kalupahana (*Nāgārjuna: The Philosophy of the Middle Way,* 1986) have been either completely omitted or alluded to only casually. Those omissions will undoubtedly disturb some readers who are familiar with Mādhyamika studies and hold that one or another of these works is essential to *any* discussion of Nāgārjuna or Buddhist studies. But the desire to include all relevant works was subsidiary to the issues of interpretive influence, assumption, and bias that are the real focus of this work. Thus, each of the works of scholars that I have discussed—Stcherbatsky, Murti, Potter, Matilal, Streng, and others—stands as a major representative of a *type* of scholarship or as indicative of a significant methodological shift. Again, the focus of the work is not an attempt to critically understand one of the central works of Buddhist thought, but rather an examination of the role that modern, Western philosophy plays in the interpretation of non-Western classic texts. This has been a study in cross-cultural hermeneutics.

Similarly, I have been careful to avoid advancing a "new" interpretation of Nāgārjuna's thought and I nowhere argue that any of the interpretations under scrutiny are "inaccurate" or "wrong." Lastly, I am interested neither in using Indian philosophical apparatus to solve Western philosophical problems nor in using Western philosophical language to redescribe ancient Indian philosophical concerns.

What *is* of interest to me is usually found in introductions or afterwords to books that do any or all of these tasks that I have tried to avoid. It is in the nonessential appendages to the text that scholars tend to discuss their agenda and objectives, their own formative influences, and their admiration for related works. The introduction and the conclusion are where authors identify with an intellectual tradition and a scholarly orientation, and where they explain why they wanted to write just this kind of book. It is widely held that these formal, subjective, and contextual concerns are of only marginal importance when compared with the basis of textual scholarship: translation and exegesis. Consequently, these marginal concerns are normally confined to the preface and the afterword. But it is precisely this category of the scholarly *hors d'oeuvre* and dessert—issues of style, influence, fashion, and presupposition—that I have wanted to examine.

It is a contemporary belief that the subject of scholarly self-consciousness is of central importance among proponents of interpretive movements such as "deconstruction," "hermeneutics," and "critical theory." There are innumerable ways in which any scholar's work will reflect ideological, professional, cultural, and psychological determinants that cannot be known without self-reflection, and any number of reasons why awareness of their own assumptions will be advantageous to scholars. As students are slowly inducted into what Thomas Kuhn has called a "paradigm" or "disciplinary matrix,"[1] they are taught what kinds of problems are professionally interesting and worthwhile to consider, and which things in their field of scholarly vision can be ignored as trivial or irrelevant to their course of study. They are also "taught" at what point their investigations should begin (what truths may be considered axiomatic or theoretically primitive) and at what point they should end (what will constitute a convincing demonstration of their thesis or an adequate exploration of the subject matter). Every scholarly community—or what Stanley Fish calls an "interpretive community"[2]—will have its own "symbolic generalizations, models, and exemplars"[3] and every generation of scholars will have its own methodological trends, stylistic preferences, and lists of canonical texts and founding fathers. In addition, influencing any scholarly enterprise will be social, political, and economic developments, religious and national loyalties, and innumerable unconscious cultural determinants—all the stuff of a *Zeitgeist*. Frank Kermode has put this succinctly:

> our period discourse is controlled by certain unconcious constraints, which make it possible for us to think in some ways to the exclusion of others. However subtle we may be at reconstructing the constraints of past [or foreign] *epistèmes,* we cannot ordinarily move outside the tacit system of our own.[4]

A product of scholarly labor will at some level also be a reflection of the author's psychology and personal history, and although currently very much out of favor in many scholarly circles,[5] consideration of these factors was seen as ultimately important by, for example, William James:

> The history of philosophy is to a great extent that of a certain clash of human temperaments. . . . Of whatever temperament a professional philosopher is, he tries, when philosophizing, to sink the fact of his temperament. Temperament is no conventionally recognized reason, so he urges impersonal reasons only for his conclusions. Yet his temperament really gives him a stronger bias than any of his more strictly objective premises. It loads the evidence for him one way or the other, making for a more sentimental or a more hard-hearted view of the universe, just as this fact or that principle would. He *trusts* his temperament. Wanting a universe that suits it, he believes in any representation of the universe that does suit it.[6]

It is impossible to discard all professional, cultural, and psychological determinants—impossible, and strictly meaningless––but one can be self-conscious about them. Textual interpretation may at times seem a very specialized activity, but it should certainly be governed by common sense. This means that excess should be avoided, and excess may come in a variety of guises. Acceptance of the insight that our minds are conditioned to see the world in certain ways and not in others has engendered at least two kinds of extreme views. One proceeds from this inference to the conclusion that we are irrevocably confined to linguistic and cultural communities, and that real understanding cannot exist among cultures, historical periods, or even individuals. This rather extreme version of self-consciousness is often termed "relativism" and although few writers actually claim this outlook for themselves—if one really believed in the impossibility of communication, then writing and publishing would hold little attraction, I suppose—many theorists in philosophy and the social sciences have accused one another of going too far in this direction, of being self-conscious to the point of irrationality.[7] The second stance that proceeds from self-consciousness about our intellectual limits leads not to relativistic despair but to intellectual license. If our beliefs, our theories, and our ways of communicating them can be reduced to questions of rhetorical style, fashion, and habit, then, one might reason, we are free to discard the constraints of these conventions and say whatever we like. We are free, moreover, to say it in whatever way we choose, our only aim being to shock our readers into an orgy of what Harold Bloom calls "perversity." Certain practitioners of deconstruction in philosophy and literary criticism write with the express purpose of liberating themselves and their readers from previously unexamined habits of "logocentric bias" and "interpretive method."[8] In self-consciousness they seem to have discovered a road to freedom, where relativists have discovered a different metaphor: nothing but an endless row of locked doors and shuttered windows.

The most common type of excess is unconscious, and that is precisely the point. Lack of self-consciousness is common in all kinds of scholarship and there is no shortage of authors who seem to feel that their particular interpretive standpoint is absolute: that they are answering universal questions in serene, unconditioned objectivity. There is an illusion of scholarly omnipotence or unself-consciousness to be found in any attempt to locate "foundational truth," to see "knowledge" as a correct "representation of reality," to find "transcendental" or "scientifically objective" methodologies, to engage in logical, linguistic, structural, phenomenological, or conceptual analyses that deny the social context of belief. Just as extreme as the relativists and deconstructionists are the ordinary scholars who believe in unconditioned facts and objective readings of texts: the unwillingness to question presuppositions is as much a failure of moderation as either the paralysis or license that can come from too much self-consciousness.

If there is a problem with viewing scholarship in this manner, my manner, it is that the scholar's pride in his achievement could decline, for moderation is never exciting. Scholars are often motivated by the belief that they are simply pursuing objective truth; but this project is necessarily complicated by the phenomenon, universally acknowledged, but rarely taken into account, that I have been calling *isogesis*. The isogetical aspect of interpretation, which Gadamer terms the "productive attitude," is a reflection of the degree to which readers see what they want to see or have been trained to see in any interpreted text. And this idea of isogetical reading, which implies that objectivity in textual interpretation is necessarily illusory, also carries with it implications about the continuity of interpretive traditions. Following the lead of thinkers such as Kuhn, Hacking, Gadamer, and Foucault, who tend toward charting the discontinuities—the "revolutions" or "theoretical shifts"—in intellectual history, I have suggested that there are epistemic breaks between successive phases of understanding. But the issue of how radical these breaks are, whether these types of understanding are "incommensurable" in some way, is little more than a red herring. Clearly, communication takes place among scholars who work from the radically different perspectives we have discussed. Idealist and Wittgensteinian interpretations of Nāgārjuna are being written at the same time and arguments are documented in the literature. In the same way, the analytic historical moment is over for some and not for others.

As scholarship is not exempt from the trends it seeks to trace,[9] it must also be that my own attempt to understand a tradition of interpretation is itself a part of that tradition. I am writing at a time when hermeneutic self-consciousness matters, and so my method may appear, as it does to me, to resemble "common-sense." But there is, of course, no way to exempt my own writing, even the present methodological remarks, from these considerations. The degree to which I have been philosophically influenced by Wittgenstein, Eliot, Quine, Rorty, Kuhn, Gadamer, Geertz, and Foucault should be clear. My own readings of the Indian sources are also heavily influenced.

However, as I have said, the suspicion that one's own investigations are neither objective nor timeless is no cause for panic. Recognition of the isogetical nature of interpretation neither invalidates work nor offers interpretive license. To paraphrase T. S. Eliot, the existing traditions of accumulated texts and readings and rereadings of texts are, in fact, "that which we know."[10] Consequently, the most we can hope for from our own interpretation is to provide, in Richard Rorty's words, "the culminating reinterpretation of our predecessors' reinterpretation of their predecessors' reinterpretation."[11] And this view, that our understanding of a subject or a text is conditioned, theory-laden, and somehow "of the moment," does not invite fears of any radical kind of discontinuity into the conversation. Every contribution to the interpretive tradition does count, however pragmatically modified a notion of "progress" we might have. As Clifford Geertz writes:

> Studies do build on other studies, not in the sense that they take up where the others leave off, but in the sense that, better informed and better conceptualized, they plunge more deeply into the same things.[12]

Prediction is not a necessary part of the humanistic scholarly process, but it has not been difficult to anticipate current studies trumpeting the arrival of a new deconstructionist interpretation of Mādhyamika (the already mentioned Magliola work is a prominent example).[13] More parallels between Derrida and Nāgārjuna are clearly waiting to be written as the contemporary enthusiasm for the new "nonmethod" gains adherents among Asianists. When Richard Rorty writes about Derrida, for instance, he observes that Derrida "has no interest in bringing 'his philosophy' into accord with common sense. He is not writing a philosophy. He is not giving an account of anything; he is not offering a comprehensive view of anything. He is not protesting against the errors of a philosophical school."[14] These are descriptions that will be hard to resist for a young Buddhologist weaned on the Wittgensteinian reading of Nāgārjuna and ready to move on to a "clearer" understanding of the text. And when that happens it should invite further studies like this one, and may inspire a retelling of the entire story.

In conclusion, it may still be necessary to guard against certain possible readings of this text and state further what it is that I do not want to say. The survey of historically placed trends that make up the body of this work is not intended to suggest that every previous attempt at cross-cultural philosophical study has failed—either because spurious comparisons were drawn, or because every previous interpretation has been fatally conditioned by some outmoded Western concern. This study offers neither an indictment of the comparative enterprise nor a suggestion that any new interpretive perspective—such as post-Wittgensteinian/Heideggerian holism, Whiteheadian process-philosophy, Gadamerian hermeneutics, Derridian grammatology, Rortian neopragmatism, or some other, as yet unrevealed interpretive mode—will offer anything like a "final and infallible,"[15] or even a wholly satisfying, interpretation.[16] A new interpretation is often a response to a new set of critical concerns, not a solution to a formerly unsolved or poorly solved puzzle. Looked at in this way, the post-Wittgensteinian fondness for Mādhyamika Buddhism is no less determined by contemporary trends than was the late-nineteenth-century fascination with Advaita Vedānta. What we do seem to have is a collection of intelligent misreadings, and that may be enough.

My point is perhaps better made by quoting a question that Hugh Kenner posed in *The Pound Era:*

> Is the life of the mind a history of interesting mistakes? More pertinently, is the surest way to a fructive Western idea the misunderstanding of an Eastern one?[17]

We should not be surprised that interpretation is not an exact science. After all, translation is not an exact science. Science is not an exact science.

NOTES

Chapter 1

1. Garland Cannon, *Oriental Jones* (London: Asia Publishing House, 1964), p. 141.

2. Edward Said, *Orientalism* (New York: Random House, 1978), p. 78.

3. Guy R. Welbon, *The Buddhist Nirvāṇa and Its Western Interpreters* (Chicago: University of Chicago Press, 1968), p. 22.

4. Jeannette Mirsky, *Sir Aurel Stein* (Chicago: University of Chicago Press, 1977), p. 18.

5. A. J. Arberry, "The Founder: William Jones," *Oriental Essays: Portraits of Seven Scholars* (London: George Allen & Unwin, 1960), pp. 48–86.

6. Maurice Winternitz, *History of Indian Literature*, 2 vols. (New Delhi: Munshiram Manoharlal, 1927), vol. 1, p. 9.

7. Ibid., p. 11.

8. Friedrich Schlegel, as quoted in Winternitz, *History of Indian Literature*, vol. 1, p. 14.

9. A. Leslie Willson, *A Mythical Image: The Ideal of India in German Romanticism* (Durham, N.C.: Duke University Press, 1964), pp. 113–114.

10. Winternitz, *History of Indian Literature*, vol. 1, p. 19.

11. Arthur Schopenhauer, *The World as Will and Representation*, trans. E. F. J. Payne, 2 vols. (New York: Dover Press, 1966), vol. 1, p. xv.

12. Arthur Schopenhauer, *Samtliche Werke*, ed. Eduard Grisebach, 6 vols. (Leipzig: Brockhaus, 1921–24), vol. 5, p. 418.

13. Ibid., vol. 2, p. 558.

14. Schopenhauer, *The World as Will and Representation*, vol. 1, pp. xv–xvi.

15. Said, *Orientalism*, p. 2.

16. Monier Williams, *Indian Wisdom* (Varanasi: Chowkhamba Sanskrit Series Office, 1963), p. 3.

17. David E. Linge, ed., *Philosophical Hermeneutics,* by Hans-Georg Gadamer (Berkeley: University of California Press, 1976), p. xiv.

18. Welbon, *The Buddhist Nirvāṇa,* p. 300.

19. Clifford Geertz borrowed this term from Gilbert Ryle and extended its use. See Clifford Geertz, "Thick Description: Toward an Interpretive Theory of Culture," *The Interpretation of Cultures* (New York: Basic Books, 1973), pp. 3–30.

20. Richard H. Robinson, "Comments on 'Non-Western Studies: The Religious Approach' by Wilfred Cantwell Smith," *A Report on an Invitational Conference on the Study of Religion in the State University* [New Haven: The Society for Religion in Higher Education, nd. (Conference held Oct. 23–25, 1964)], p. 64.

21. Surendranath Dasgupta's five-volume *History of Indian Philosophy,* Erich Frauwallner's two-volume *History of Indian Philosophy,* Jadunath Sinha's two-volume *History of Indian Philosophy,* M. Hiriyanna's *Outlines of Indian Philosophy,* A. K. Warder's *Outline of Indian Philosophy*—none of these works is actually "historical" in approach.

22. Linge, ed., *Philosophical Hermeneutics,* p. xiv.

23. T. S. Eliot, *Christianity and Culture* (New York: Harcourt, Brace & World, 1949), p. 115.

24. Welbon, *The Buddhist Nirvāṇa,* p. 298.

25. Jeffrey Stout, "What Is the Meaning of a Text?," *New Literary History,* 14 (1982–83), p. 9.

26. Richard Rorty, *Consequences of Pragmatism* (Minneapolis: University of Minnesota Press, 1982), p. 140.

27. Hans-Georg Gadamer, *Truth and Method* (New York: Crossroad, 1975), p. 263.

28. Stout, "What Is the Meaning of a Text?," p. 8.

29. Richard Rorty, *Philosophy and the Mirror of Nature* (Princeton: Princeton University Press, 1979), p. 264.

30. Ibid., pp. 131–132.

31. Paul Deussen, *The Philosophy of the Upanishads,* trans. A. S. Geden (Edinburgh: T. & T. Clark, 1906), pp. 40–41.

32. Ibid., p. 45.

33. Ibid., p. 46.

34. Williams, *Indian Wisdom,* p. 61.

35. The traditional Brahmin system of classifying Indian philosophy into six separate *darśanas* or viewpoints has always had problems of incompleteness and inconsistency. The omission of the entire range of *nāstika* schools is clearly the most serious difficulty. Not only are Buddhist, Jain, Cārvāka, and Ājīvika

teachings excluded from consideration, but their significant influence on *āstika* systems is also ignored. A large part of the polemical material in Śaṃkara's *Brahmasūtrabāṣya* is devoted to refuting the Buddhists. Karl Potter notes that "apparently *he* did not consider them refuted by the mere fact of their being outside the fold." [See Potter, *Presuppositions of India's Philosophies* (Englewood Cliffs, N.J.: Prentice-Hall, 1963), pp. 98–102.] Śaṃkara also incorporated Buddhist ideas, terms, and dialectical style into his own texts—his term, *adhyāsa,* "superimposition," is Buddhistic and his use of Nāgārjuna's *catuṣkoti,* or "four-cornered negation," is a clear instance of influence. Furthermore, other "orthodox" philosophers incorporated "heterodox" views. Kumārila accepts numerous Jain assumptions, and Prabhākara, as well as several Advaitins employ the Buddhist doctrine of *apohavāda* (the analysis of similarity as the "nonintuiting of difference"). Gaudapada, the author of the *Māṇḍūkyakārikās,* was both an Advaitin and a Buddhist. Within the study of the *āstika* traditions, the "six-schools" systematization faces other difficulties. It separates philosophers whose views are almost identical and combines opposed philosophies into one school. The clearest example of this is the classification of the views of both Prabhākara and Kumārila-bhaṭṭa under the label of Mīmāṃsaka. There are historical reasons why they are placed together, but their views diverge on almost all technical issues. Kumārila's list of categories is much more Nyāya-like than Prabhākara's, as it recognizes negative categories. Their conceptions of the nature of the *ātman* differ; Kumārila states that it is the object of self-consciousness and Prabhākara states that it is the subject. They also disagree in regard to the nature of illusion and in their philosophies of action. In fact, they are so far apart in most of their positions that accounts of the Mīmāṃsā school continually shift between the two eighth-century philosophers. [See P. T. Raju, *The Philosophical Traditions of India* (Pittsburgh: University of Pittsburgh Press, 1971), pp. 66–85.] Another serious oversimplification occurs under the label of Vedānta. The term is commonly used to refer to Advaita, Dvaita, Viśiṣṭadvaita, and as many as 12 other philosophical systems. There is more divergence, for example, between Śaṃkara's Advaita and Madhva's Dvaita than between Dvaita and Nyāya. The term "Vedānta" is no more exact or descriptive a label for a philosophical position than is "Buddhism." Likewise, the traditional practice of separating the Nyāya of Gautama and the Vaiśeṣika of Kaṇāda is philosophically problematic. The two views rarely are in opposition on any major issue and actually join forces quite early in their histories. The Sāṃkhya and Yoga schools supplement each other in a similar way.

36. The following classic demonstration of the "Indian" syllogism is *always* included in chapters on the Nyāya school:

1. This mountain has fire;
2. because it has smoke;

3. wherever there is smoke, there is fire, as in the kitchen,
4. this mountain also has smoke that is acompanied by fire;
5. thus, this mountain has fire.

37. Arthur Berriedale Keith, *Indian Logic and Atomism* (Oxford: Clarendon Press, 1921).

38. Keith's *Indian Logic and Atomism* was the first major Western work to be devoted exclusively to Nyaya-Vaisesika. Richard Garbe's *Die Sāṃkhya Philosophie* (Leipzig: H. Haessel, 1897) was the first devoted to Sāṃkhya. Two earlier nineteenth-century works had "Sāṃkhya" in their titles but present only sketchy expositions of Sāṃkhya materials: E. Roer, *Lecture on the Sāṃkhya Personality* (Calcutta: np, 1854), C. B. Schluter, *Aristotles' Metaphysik eine Tochter der Sāṃkhya-Lehre des Kapila,* (Munster: np, 1874).

39. Keith, *Indian Logic,* p. 3.

40. Max Muller, *The Six Systems of Indian Philosophy* (Varanasi: Chowkhamba Sanskrit Series Office, 1899), p. 362.

41. Ibid., p. 196.

42. Ibid., p. 197–198.

43. Ibid., p. v.

44. Paul Deussen, "The Philosophy of the Vedānta in Its Relations to the Occidental Metaphysics," *Journal of the Asiatic Society of Bengal,* 13 (1890), pp. 330–340. Reprinted Bombay, 1893.

45. Brahma Sūtras, IV, i, 3.

46. Eliot Deutsch, *Advaita Vedānta: A Philosophical Reconstruction* (Honolulu: East-West Center Press, 1969), p. 27.

47. Paul Deussen, *Outline of the Vedānta* (London: Luzac & Co., 1907), p. vii.

48. See Karl Potter, "A Fresh Classification of India's Philosophical Systems," *Journal of Asian Studies* 21:1 (November 1961), pp. 25–32; Ninian Smart, *Doctrine and Argument in Indian Philosophy* (London: George Allen & Unwin, 1964); and P. T. Raju, *Philosophical Traditions of India* for criticisms of and alternatives to the traditional classification strategies.

49. S. N. L. Shrivastava, *Śaṃkara and Bradley* (Delhi: Motilal Banarsidass, 1968), pp. 2–3.

50. Walter Ruben, introduction to Debiprasad Chattopadhyaya, *Indian Philosophy, A Popular Introduction* (New Delhi: People's Publishing House, 1964), p. xix.

51. Chattopadhyaya, *Indian Philosophy,* pp. xiii–xiv.

52. See D. R. Shastri, *A Short History of Indian Materialism* (Calcutta: University of Calcutta, 1930); Jadunath Sinha, *Indian Realism* (London: Routledge & Kegan Paul, 1938); Erich Frauwallner, *History of Indian Philosophy,* vol. 2 (Delhi: Motilal Banarsidass, 1956); A. K. Warder, *Outline of Indian*

Philosophy (Delhi: Motilal Banarsidass, 1971); Walter Ruben, *Geschichte der Indichen Philosophie* (Berlin: Deutscher Verlag der Wissenschaften, 1954); Dale Riepe, *The Naturalistic Tradition in Indian Thought* (Seattle: University of Washington Press, 1961); K. K. Mittal, *Materialism in Indian Thought* (Delhi: Munshiram Manoharlal, 1974); and D. D. Mehta, *Positive Sciences in the Vedas* (Delhi: Arnold Heineman Publishers, 1974).

53. See A. L. Basham, *History and Doctrines of the Ājīvikas: A Vanished Indian Religion* (London: Luzac & Co., 1951), and Debiprasad Chattopadhyaya, *Lokāyata, A Study in Ancient Indian Materialism* (Calcutta: University of Calcutta, 1959).

54. See D. H. H. Ingalls, *Materials for the Study of Navya-nyāya Logic* (Cambridge: Harvard University Press, 1951); Karl Potter, *Presuppositions of India's Philosophies* (Westport, Conn.: Greenwood Press, 1963); R. C. Pandeya, *The Problem of Meaning in Indian Philosophy* (Delhi: Motilal Banarsidass, 1963); D. N. Shastri, *Critique of Indian Realism* (Agra: Agra University, 1964); B. K. Matilal, *The Navya-nyāya Doctrine of Negation* (Cambridge: Harvard University Press, 1968); Dhirendra Sharma, *The Differentiation Theory of Meaning in Indian Logic* (The Hague: Mouton, 1969); B. K. Matilal, *Epistemology, Logic, and Grammar in Indian Philosophical Analysis* (The Hague: Mouton, 1971); R. R. Dravid, *The Problem of Universals in Indian Philosophy* (Delhi: Motilal Banarsidass, 1972); M. C. Bhartiya, *Causation in Indian Philosophy* (Ghaziabad: Vimal Prakashan, 1973); and B. K. Matilal, *Nyāya-Vaiśeṣika* (Wiesbaden, Germany: Otto Harrassowitz, 1977).

55. Robert Thurman, "Philosophical Nonegocentrism in Wittgenstein and Candrakīrti in their Treatment of the Private Language Problem," *Philosophy East and West,* 30:3 (July 1980), p. 336.

56. Chris Gudmunsen, *Wittgenstein and Buddhism* (New York: Harper & Row, 1977), p. 115.

57. Ibid., p. 113.

58. Ibid., p. viii.

59. Frederick Streng, *Emptiness* (Nashville: Abingdon Press, 1967), p. 139.

60. See Mervyn Sprung, ed., *The Problem of Two Truths in Buddhism and Vedānta* (Dordrecht, Holland: D. Reidel, 1974).

Chapter 2

1. Engelbert Kaempfer, *The History of Japan Together with a Description of the Kingdom of Siam, 1690–1692,* trans. J. G. Scheuchaer (Glasgow: James MacLehose and Sons, 1906), vol. 2, p. 56.

2. A. J. Arberry, "The Founder: William Jones, "*Oriental Essays: Portraits of Seven Scholars* (London: George Allen & Unwin, 1960), p. 80.

3. Guy Richard Welbon, *The Buddhist Nirvāṇa and Its Western Interpreters* (Chicago: University of Chicago Press, 1968), p. 28. See also J. W. de Jong, "Brief History of Buddhist Studies in Europe and America," *The Eastern Buddhist* 7 (1974) for another, more recent, comprehensive survey of the scholarly history.

4. Brian Houghton Hodgson, *Essays on the Languages, Literature, and Religion of Nepal and Tibet* (London: Williams & Norgate, 1874), p. 98.

5. Eugène Burnouf, *Papiers d' Eugène Burnouf conservés à la Bibliothèque Nationale* (Paris: np, 1899), translated in Welbon, *Nirvāṇa,* p. 62.

6. William Peiris, *The Western Contribution to Buddhism* (Delhi: Motilal Banarsidass, 1973), p. 168.

7. Welbon, *Nirvāṇa,* p. 68.

8. Jules Barthelemy Saint-Hilaire, *Le Bouddha et sa religion* (Paris: Didier, Nouv. ed., 1862), pp. xxix–xxx.

9. Welbon, *Nirvāṇa,* p. 73.

10. Saint-Hilaire, *Le Bouddha,* p. viii.

11. Welbon, *Nirvāṇa,* p. 88.

12. Ibid., p. 91.

13. Welbon, *Nirvāṇa,* p. 76.

14. The term, "nihilism," came into vogue, borrowed from popular Russian writers such as Turgenev and Tolstoy in the second-half of the nineteenth century—at the same time that Buddhist texts were first being considered seriously. It may be that Buddhism was heralded as a prime example of this new, fashionable brand of extremist philosophy.

15. Louis de La Vallée Poussin, *Bouddhisme: Opinions sur l'histoire de la dogmatique* (Paris: Gabriel Beauchesne et Cie., 1908), p. 101.

16. Nāgārjuna's text is not itself titled, but is often referred to either as the *Mādhyamikakārikās* (Verses on the Middle Way), the *Mādhyamikaśāstra* (Treatise on the Middle Way), or the *Mūlamādhyamikakārikās* (Verses on the Principles of the Middle Way). The spelling, "Mādhyamaka," is a sometimes-used variant of "Mādhyamika" and, according to Murti, may have been an earlier term used for the Middle Way. [See T. R. V. Murti, *The Central Philosophy of Buddhism* (London: George Allen & Unwin, 1955), p. 87.]

17. Fyodor Stcherbatsky, "Review of *Nirvāṇa* by Louis de La Vallée Poussin," *Bulletin of the School of Oriental Studies,* 4, (1926), pp. 359–360.

18. Welbon, *Nirvāṇa,* p. 290.

19. Fyodor Stcherbatsky, *Buddhist Logic,* 2 vols. (Leningrad: Academy of Sciences of the U.S.S.R., 1930), 1:7.

20. Immanuel Kant, *Critique of Pure Reason,* trans. Norman Kemp Smith (New York: St. Martin's Press, 1929), p.93.

21. Stcherbatsky, *Buddhist Logic,* vols. 1 and 2.

22. Ibid., 1:73.
23. Ibid.
24. Ibid., p. 212.
25. Ibid., p. 84.
26. Ibid., p. 103.
27. Ibid., p. 105.
28. Fyodor Stcherbatsky, *The Central Conception of Buddhism* (London: Royal Asiatic Society, 1923), pp. 37–38.
29. Ibid., p. 62.
30. Richard H. Robinson, "Comments on 'Non-Western Studies: The Religious Approach' by Wilfred Cantwell Smith," A Report on an Invitational Conference on the Study of Religion in the State University [New Haven: The Society for Religion in Higher Education, nd. (Conference held October 23–25, 1964)], p. 63.
31. Stcherbatsky, *Buddhist Logic*, 1:12.
32. Ibid., p. 14.
33. *Mādhyamikakārikā*, Chapters I–XXI. In most cases, the translations used throughout this study follow Kenneth K. Inada, *Nāgārjuna, A Translation of his Mūlamadhyamaka-kārikā* (Tokyo: The Hokuseido Press, 1970), with occasional modifications.
34. Mādhyamikakārikā, Chapter 25.
35. See Stanislaw Schayer's *Ausgewahlte Kapitel aus der Prasannapadā* (Krakow: np, 1931).
36. Fyodor Stcherbatsky, *The Conception of Buddhist Nirvāṇa* (Delhi: Motilal Banarsidass, 1968), p. 45.
37. Ibid., p. 38.
38. Stcherbatsky, *Buddhist Logic*, 1:9.
39. Ibid., pp. 9–10.
40. Stcherbatsky, *Conception of Buddhist Nirvāṇa*, p. 61.
41. Ibid., p. 60.
42. T. R. V. Murti, *The Central Philosophy of Buddhism* (London: George Allen & Unwin, 1955), p. 234.
43. Ibid., p. 5.
44. Karl Potter, *Presuppositions of India's Philosophies* (Westport: Greenwood Press, 1963), p. 239.
45. Murti, *Central Philosophy of Buddhism*, p. 209.
46. Ibid., p. 280.
47. Ibid., pp. 321–322.
48. Ibid., p. 217.
49. Ibid., p. 237.
50. Ibid.

51. Murti, "Saṃvṛti and Paramārtha," in *The Problem of Two Truths in Buddhism and Vedānta,* ed. Mervyn Sprung (Dordrecht: D. Reidel, 1973), p. 17.

52. Ibid., p. 19.

53. Ibid., p. 22.

54. Ibid., p. 9.

55. Ibid., p. 22.

56. Ibid.

57. Murti, *Central Philosophy of Buddhism*, p. 123.

58. Ibid., p. 320.

59. Ibid., p. 327.

60. Ibid., p. 330.

Chapter 3

1. Although there was quite a significant time lag between the development of the new trends in philosophy proper and the new style among the Asianists (Russell's "On Denoting," an analytic milestone, had already been published in 1905), this pattern followed the precedent set by the idealist interpreters of the previous century. In both cases it took several decades for new Western philosophical ideas to become widespread enough to be accepted as presuppositional by students of the Asian traditions.

2. See Alex Wayman, "Who Understands the Four Alternatives of the Buddhist Texts?" *Philosophy East and West,* 27:1 (January 1977), pp. 3–21.

3. Richard H. Robinson, "Some Logical Aspects of Nāgārjuna's System," *Philosophy East and West,* 6:4 (January 1957), pp. 291–292.

4. Ibid., p. 292.

5. Ibid.

6. Ibid.

7. Ibid., p. 293.

8. Ibid., p. 294.

9. Ibid., pp. 302–303.

10. Ibid., p. 305.

11. Ibid., p. 307.

12. Robinson, "Classical Indian Philosophy," *Chapters In Indian Civilization,* ed. Joseph W. Elder (Dubuque, Iowa: Kendall/Hunt, 1970), p. 206.

13. *Mādhyamikakārikā,* Chapter XV, verse 2.

14. Robinson, *Early Mādhyamika in India and China* (Madison: University of Wisconsin Press, 1967), p. 41.

15. Robinson, "Classical Indian Philosophy," pp. 206–207.

16. Ibid., p. 207.

17. Ibid.
18. Ibid.
19. Ibid., p. 208.
20. Ibid., p. 205.
21. Ibid., p. 206.
22. Ibid.
23. Ibid., p. 207.
24. Ibid., p. 208.
25. Ibid., p. 210.
26. Robinson, *Early Mādhyamika*, pp. 57–58.
27. Although *Early Mādhyamika in India and China* is a full, book-length study, most of the work is devoted to the translation and analysis of fourth-century Chinese texts. The 50 pages actually concerned with Indian Mādhyamika are composed largely of bibliographic materials. There are only very brief sections devoted to a discussion of Nāgārjuna's philosophy. The writing in these sections (no more than 20 pages) is so terse as to be practically epigrammatic.
28. Karl H. Potter, *Presuppositions of India's Philosophies* (Englewood Cliffs, N.J.: Prentice-Hall, 1963), p. 3.
29. Orientalists of the nineteenth and early-twentieth centuries were virtually unanimous in proclaiming the lack of concern for "morality" in Indian texts: A. B. Keith wrote that "in comparison with the intellectual activities of the Brahmins the ethical content of the Upanishads must be said to be negligible and valueless," and R. E. Hume claimed that "the possession of metaphysical knowledge actually cancels all past sins and even permits the knower unblushingly to continue in 'what seems to be much evil' with perfect impunity." As late as 1950, Albert Schweitzer was denouncing Indian thought as "world-and-life denying."
30. Although Potter does not name his Western philosophical sources, he often clearly echoes Moritz Schlick's prescription that "desire for the truth is the only appropriate inspiration for the thinker when he philosophizes; otherwise his thoughts run the danger of being led astray by his feelings. His wishes, hopes, and fears threaten to encroach upon that objectivity which is the necessary presupposition of all honest inquiry." [Schlick, "What Is the Aim of Ethics," in *Logical Positivism*, ed. A. J. Ayer (New York: The Free Press, 1959), p. 247.] Potter's treatise is thoroughly consistent with the viewpoint of the popular bible of positivism, Ayer's *Language, Truth and Logic*.
31. Potter, *Presuppositions*, p. 12.
32. Ibid., p. 11.
33. Ibid., p. 12.
34. See B. F. Skinner, *Science and Human Behavior* (New York: The Free Press, 1953) for one of the seminal expositions of behaviorist terminology.

35. Potter, *Presuppositions,* p. 12.

36. Ibid., p. 13.

37. Ibid., p. 16.

38. Ibid., p. 23.

39. Ibid., p. 103. The major distinction among speculative philosophers in Potter's system is that between what he calls "progress" philosophers and "leap" philosophers. A progress philosopher believes that there is a chain of possible events that leads to *mokṣa,* and that the events in the chain are connected by some relation that is neither so strong that the events are out of the agent's control, nor so weak that the agent cannot depend on one kind of event following its causal precedent. A leap philosopher denies that a determined chain of events can include *mokṣa* or that *mokṣa* is an event. Also included within the "leap" category are those who assert that God's grace is one of the necessary links in the chain (for example, Madhva's Dvaita).

On his chart, Potter names two leap philosophies, eleven progress philosophies, one skeptical school, and one fatalist school. The progress philosophies are divided into three types of views on the relations between causes and effects, and all schools are further divided by their views on five other issues such as the nature of illusion, and the relation between the whole and its parts. All philosophical questions, however, are seen as subsidiary to questions of causality.

40. Ibid., p. 99.

41. Ibid., p. 238.

42. Ibid., p. 241.

43. Ibid., p. 240.

44. Ibid., p. 239.

45. Ibid., p. 114.

46. Ibid., p. 241.

47. Bimal K. Matilal, *Epistemology, Logic, and Grammar in Indian Philosophical Analysis* (The Hague, The Netherlands: Mouton, 1971), p. 166.

Chapter 4

1. Richard H. Robinson, *Early Mādhyamika in India and China* (Madison: University of Wisconsin Press, 1967), p. 49.

2. Bimal K. Matilal, *Epistemology, Logic, and Grammar in Indian Philosophical Analysis* (The Hague, The Netherlands: Mouton, 1971), p. 167.

3. Ives Waldo, "Nāgārjuna and Analytic Philosophy, II," *Philosophy East and West,* 28:3 (July 1978), pp. 295–296.

4. W. V. O. Quine, *From a Logical Point of View* (New York: Harper & Row, 1953), p. 42.

5. Donald Davidson, "Truth and Meaning," *Synthese,* 18 (1967), pp. 304–323.

6. Nathan Katz, "Nāgārjuna and Wittgenstein on Error," in *Buddhist and Western Philosophy,* ed. Nathan Katz (New Delhi: Sterling, 1981), p. 311.

7. Chris Gudmunsen, *Wittgenstein & Buddhism* (London: Macmillan, 1977), p. 67.

8. Katz, "Nāgārjuna and Wittgenstein," p. 312.

9. Ludwig Wittgenstein, *Philosophical Investigations,* trans. G. E. M. Anscombe (New York: Macmillan, 1958), p. 48e.

10. Ibid., p. 103e.

11. Ibid., p. 47e.

12. Wittgenstein, *The Blue and Brown Books* (New York: Harper, 1958), p. 27.

13. Katz, "Nāgārjuna and Wittgenstein," p. 313.

14. Frederick Streng, *Emptiness* (Nashville: Abingdon Press, 1967), pp. 139–141.

15. Ibid., p. 142.

16. Gudmunsen, *Wittgenstein & Buddhism,* p. 68.

17. See Robert A. F. Thurman, *Tsong Khapa's Speech of Gold in the Essence of True Eloquence,* (Princeton: Princeton University Press, 1984).

18. Streng, *Emptiness,* p. 80.

19. Katz, "Nāgārjuna and Wittgenstein," p. 306.

20. Ibid.

21. Ibid., p. 318.

22. Wittgenstein, *Philosophical Investigations,* p. 48e.

23. Katz, "Nāgārjuna and Wittgenstein," pp. 319–320.

24. Murti had cited this kind of Buddhistic position as an analogue to the Upaniṣadic "neti, neti"—a way of expressing the "indescribability of the absolute." [Murti, *The Central Philosophy of Buddhism* (London: George Allen & Unwin, 1960), p. 48]

25. Streng, *Emptiness,* p. 80.

26. Tsong Khapa, *Speech of Gold,* p. 122.

27. Ibid, p. 123.

28. Thurman, "Philosophical Nonegocentrism in Wittgenstein and Candrakīrti in their Treatment of the Private Language Problem," *Philosophy East and West,* 30:3 (July 1980), pp. 326–327.

29. Katz, "Nāgārjuna and Wittgenstein," p. 319.

30. See Matilal, *Epistemology, Logic, and Grammar,* p. 42.

31. Katz, "Nāgārjuna and Wittgenstein," p. 319.

32. Thurman, "Philosophical Nonegocentrism," p. 327.

33. Ibid., pp. 327–328.

34. Streng, *Emptiness,* p. 140.

35. This quote (Wittgenstein, *Philosophical Investigations,* p. 47e), with its "Buddhistic flavor," may be the favorite of the Wittgensteinian school of Nāgārjuna criticism. It appears more frequently in the literature than any other.

36. Nāgārjuna, *Mādhyamikakārikā,* Chapter 24, verse 11.

37. Streng, *Emptiness,* p. 22.

38. Ibid., p. 23.

39. Robinson, *Early Mādhyamika,* p. 49.

40. Murti, *Central Philosophy of Buddhism,* p. 209.

41. Ibid., p. 280.

42. Matilal, "A Critique of the Mādhyamika Position," *The Problem of Two Truths in Buddhism and Vedānta,* ed. Mervyn Sprung (Dordrecht: D. Reidel, 1973), p. 63.

43. Wittgenstein, *Remarks on the Foundation of Mathematics* (New York: Macmillan, 1956), p. 157.

44. Streng, "The Significance of Pratītyasamutpāda for Understanding the Relationship between Saṃvṛti and Paramārthasatya in Nāgārjuna," *The Problem of Two Truths,* p. 36.

45. In his gloss on this verse in the *Prasannapadā,* Candrakīrti quotes a *sūtra* that reports a conversation between the Buddha and Kāśyapa:

"Indeed, Kāśyapa, it were better if one resorted to a belief in the reality of the individual as unshakable as Mount Sumeru, than to hold to a theory of the absence of being through the stubborn belief in the unreality of things. Why is that? Because Kāśyapa, the absence of being [emptiness] is the exhaustion of all theories and views."

"One for whom, in turn, the absence of being itself becomes a dogmatic view I call incurable. It is, Kāśyapa, as if a sick man were given a medicine by a doctor, but that medicine, having removed his ills, was not itself expelled but remained in the stomach. What do you think, Kāśyapa, will this man be freed of his sickness? No indeed, illustrious one, the sickness of this man in whose stomach the medicine, having removed all his ills remains and is not expelled, would be more violent. The illustrious one said: In this sense, Kāśyapa, the absence of being is the exhaustion of all dogmatic views. But the one for whom the absence of being itself becomes a fixed belief, I call incurable." [Candrakīrti, *Prasannapadā,* trans. Mervyn Sprung (Boulder: Prajña Press, 1979), pp. 150–151.]

46. Katz, "Nāgārjuna and Wittgenstein," p. 323.

47. Ibid.

48. Streng, "The Significance of Pratītyasamutpāda," p. 29.

49. Thurman, "Philosophical Nonegocentrism," p. 336.

50. Katz, "Nāgārjuna and Wittgenstein," p. 311.

Afterword

1. Thomas S. Kuhn, *The Essential Tension* (Chicago: University of Chicago press, 1977), pp. 217–298.

2. Stanley Fish, *Is There a Text in This Class? The Authority of Interpretive Communities* (Cambridge: Harvard University Press, 1980).

3. Kuhn, *Essential Tension*, p. 297.

4. Frank Kermode, *The Classic* (Cambridge: Harvard University Press, 1983), pp. 139–140.

5. The classic essay by W. K. Wimsatt and M. C. Beardsley, "The Intentional Fallacy" (1946), which is often cited as the first major statement of the New Critical doctrine that the author's intention is irrelevant to interpretation of texts, has engendered great dispute both inside and outside of literary criticism about the value of considering matters of authorial biography, psychology, and self-perception. E. D. Hirsch's well known attack on what he calls theories of "semantic autonomy" [*Validity in Interpretation* (New Haven: Yale University Press], 1967] is an unsatisfying example of the critical response. The controversies surrounding Hans-Georg Gadamer's linguistically oriented "philosophical hermeneutics" are far more fruitful.

6. William James, *Pragmatism* (New York: Longmans, Green and Co., 1907), p. 7.

7. Criticism of Kuhn's historicist view of science is a prime example of this phenomenon. Gerald Doppelt's "Kuhn's Epistemological Relativism: An Interpretation and Defense" [*Inquiry* 21 (1978), pp. 33–86] and the seventh chapter of Richard Rorty's *Philosophy and the Mirror of Nature* [(Princeton: Princeton University Press, 1979), pp. 315–356] give excellent summaries and analyses of these attacks.

8. Critical reactions to the strategies of "deconstruction" employed by Jacques Derrida, Paul de Man, Geoffrey Hartman, J. Hillis Miller, Harold Bloom, and so on, are usually those of extreme bafflement and contempt [e.g., John R. Searle, "Reiterating the differences", *Glyph,* 1 (1977), pp. 198–208] or enthusiastic support [e.g., Christopher Norris, *The Deconstructive Turn* (London: Methuen & Co.), 1983]. There has not been much written by those who are sympathetic to the aims and principles of deconstruction but who are troubled by the methods employed and the conclusions reached by deconstructionists. Richard Rorty's "Deconstruction and Circumvention" [*Critical Inquiry,* 11:1 (September 1984), pp. 1–23] is one example of a sympathetic, yet critically distant, appraisal.

9. Jeffrey M. Perl, *Skepticism and Modern Enmity: Before and After Eliot* (Baltimore: The Johns Hopkins University Press, 1989), p. 8.

10. T. S. Eliot, *Selected Essays* (New York: Harcourt, Brace & World, 1932), p. 6.

11. Richard Rorty, *Consequences of Pragmatism* (Minneapolis: University of Minnesota Press, 1982), p. 92.

12. Clifford Geertz, *The Interpretation of Cultures* (New York: Basic Books, 1973), p. 25.

13. One of the most interesting examples of the new Derridean interpretations of Nāgārjuna is to be found in Part 3 of Robert Magliola's *Derrida on the Mend* (West Lafayette, Indiana: Purdue University Press, 1986), in which Magliola argues that Nāgārjuna's Middle Way "tracks the Derridean trace, and goes 'beyond Derrida' in that it frequents the 'unheard-of-thought,' and also, 'with one and the same stroke,' allows the reinstatement of the logocentric too" (p. 87).

14. Rorty, *Consequences of Pragmatism,* p. 97.

15. Helen Gardner, *The Business of Criticism* (London: Oxford University Press, 1959), p. 51.

16. It is important to note that there are a number of other contemporary studies of Nāgārjuna offering nonabsolutist, nonrelativist, and nonanalytic readings of the Mādhyamika philosophy that do not attribute their interpretation to a "Wittgensteinian" influence. Kenneth Inada has promoted a Whiteheadian, "process"-oriented interpretation [see Inada's brief introduction to his translation of the *Mādhyamikakārikā* (previously cited) and, especially, Kenneth K. Inada and Nolan P. Jacobson, eds., *Buddhism and American Thinkers* (Albany: State University of New York Press, 1983)] that might eventually be associated with a "neopragmatist" approach. Other significant works are by David J. Kalupahana, *Nāgārjuna: The Philosophy of the Middle Way* (Albany: State University of New York Press, 1986) as well as an earlier study by Vincente Fatone, *The Philosophy of Nāgārjuna* [Delhi: Motilal Banarsidass, 1981 (originally published in 1962)]. Neither of these studies should be simplistically associated with a single, major, Western philosophical movement, but both do studiously attempt to avoid the Stcherbatskian "single-substance" ontology interpretation as well as the old-fashioned indictment of "relativism."

17. Hugh Kenner, *The Pound Era* (Berkeley: University of California Press, 1971), p. 230.

SELECTED BIBLIOGRAPHY

Anscombe, G. E. M. *Intention*. Ithaca, New York: Cornell Univ. Press, 1957.

Apel, Karl-Otto. *Analytic Philosophy of Language and the Geisteswissenschaften*. New York: Humanities Press, 1967.

Arberry, A. J. *Oriental Essays: Portraits of Seven Scholars*. London: George Allen & Unwin, 1960.

Ayer, A. J. *Language, Truth and Logic*. New York: Dover, 1952.

——— ed. *Logical Positivism*. New York: The Free Press, 1959.

——— *Philosophy in the Twentieth Century*. New York: Random House, 1982.

Banergee, Kali Krishna. "Wittgenstein v. Naiyāyika". *Calcutta Review*, 147 (1958), 27–44.

Basham, A. L. *History and Doctrines of the Ājīvikas: A Vanished Indian Religion*. London: Luzac & Co., 1951.

——— *The Wonder That Was India*. New York: Grove Press, 1954.

Baudet, Henri. *Paradise on Earth: Some Thoughts on European Images of Non-European Man*. Translated by Elizabeth Wentholt, New Haven: Yale Univ. Press, 1965.

Baynes, Kenneth, James Bohman, and Thomas McCarthy (eds.) *After Philosophy*. Cambridge: MIT Press, 1987.

Bernstein, Richard J. *Beyond Objectivism and Relativism: Science, Hermeneutics, and Praxis*. Philadelphia: Univ. of Pennsylvania Press, 1985.

——— *Philosophical Profiles*. Philadelphia: Univ. of Pennsylvania Press, 1986.

Bhartiya, M. C. *Causation in Indian Philosophy*. Ghaziabad: Vimal Prakashan, 1973.

Borger, R. and F. Cioffi (eds.). *Explanation and the Behavioural Sciences*. Cambridge: Cambridge Univ. Press, 1970.

Brown, S. C. (ed.). *Objectivity and Cultural Divergence*. Cambridge: Cambridge Univ. Press, 1984.

——— (ed.). *Philosophical Disputes in the Social Sciences*. Sussex: Harvester Press, 1979.

Candrakīrti. *Prasannapadā*. Translated by Mervyn Sprung. Boulder, Colorado: Prajña Press, 1979.

Cannon, Garland. *Oriental Jones*. London: Asia Publishing House, 1964.

Cavell, Stanley. *Must We Mean What We Say?* Cambridge: Cambridge Univ. Press, 1969.

Chattopadhyaya, Debiprasad. *Lokāyata, A Study in Ancient Indian Materialism*. Calcutta: Univ. of Calcutta, 1959.

——— *Indian Philosophy, A Popular Introduction*. New Delhi: People's Publishing House, 1964.

Conze, Edward. *Buddhism*. New York: Philosophical Library, 1951.

——— *Buddhist Thought in India*. London: George Allen & Unwin, 1962.

———"Buddhist Philosophy and Its European Parallels." *Philosophy East and West*, 13 (1963), 9–23.

——— "Spurious Parallels to Buddhist Philosophy." *Philosophy East and West*, 13 (1963), 105–115.

——— *Thirty Years of Buddhist Studies*. Oxford: Bruno Cassirer, Ltd., 1967.

Dasgupta, Surendranath. *Indian Idealism*. Cambridge: Cambridge Univ. Press, 1933.

——— *A History of Indian Philosophy*, 5 volumes. Cambridge: Cambridge Univ. Press, 1963.

Davidson, Donald. *Inquiries into Truth and Interpretation*. Oxford: Clarendon Press, 1984.

Deussen, Paul. "The Philosophy of the Vedānta in Its Relations to the Occidental Metaphysics," *Journal of the Asiatic Society of Bengal*, 13 (1890), reprinted Bombay, 1893.

——— *The Philosophy of the Upanishads*. Translated by A. S. Geden. Edinburgh: T. & T. Clark, 1906.

——— *Outline of the Vedānta*. London: Luzac & Co., 1907.

Deutsch, Eliot. *Advaita Vedānta: A Philosophical Reconstruction*. Honolulu: East-West Center Press, 1969.

Dilthey, Wilhelm. *W. Dilthey: Selected Writings*. Edited, translated, and introduced by H. P. Rickman. London: Cambridge Univ. Press, 1976.

Doppelt, Gerald. "Kuhn's Epistemological Relativism: An Interpretation and Defense." *Inquiry*, 21 (1978), 33–86.

Dravid, R. R. *The Problem of Universals in Indian Philosophy*. Delhi: Motilal Banarsidass, 1972.

Dray, William. *Laws and Explanation in History*. London: Oxford Univ. Press, 1957.

Eliot, T. S. *Selected Essays*. New York: Harcourt, Brace & World, 1932.

——— *Christianity and Culture*. New York: Harcourt, Brace & World, 1949.

Felleppa, Robert. *Convention, Translation, and Understanding: Philosophical Problems in the Comparative Study of Culture*. Albany: State University of New York Press, 1988.

Fish, Stanley. *Is There a Text in This Class? The Authority of Interpretive Communities*. Cambridge: Harvard Univ. Press, 1980.

Foucault, Michel. *The Order of Things*. New York: Random House, 1970.

——— *The Archaeology of Knowledge and the Discourse on Language*. New York: Harper & Row, 1972.

———*Language, Counter-Memory, Practice*. Ithaca, New York: Cornell Univ. Press, 1977.

Frauwallner, Erich. *History of Indian Philosophy,* 2 volumes. Delhi: Motilal Banarsidass, 1956.

Gadamer, Hans-Georg. *Truth and Method*. New York: Crossroad, 1975.

——— *Philosophical Hermeneutics*. Edited by David E. Linge. Berkeley: University of California Press, 1976.

——— *Reason in the Age of Science*. Cambridge: MIT Press, 1976.

Garbe, Richard. *Die Sāṃkhya Philosophie*. Leipzig: H. Haessel, 1897.

Gardner, Helen. *The Business of Criticism*. London: Oxford Univ. Press, 1959.

Geertz, Clifford. *The Interpretation of Cultures*. New York: Basic Books, 1973.

——— *Local Knowledge*. New York: Basic Books, 1983.

Gudmunsen, Chris. *Wittgenstein and Buddhism*. New York: Harper & Row, 1977.

Halbfass, Wilhelm. *India and Europe*. Albany: State University of New York Press, 1988.

Hacking, Ian. *Why Does Language Matter to Philosophy?* Cambridge: Cambridge Univ. Press, 1975.

Hiriyanna, M. *Outlines of Indian Philosophy*. London: George Allen & Unwin, 1932.

——— *The Essentials of Indian Philosophy*. London: George Allen & Unwin, 1949.

Hirsch, E. D. *Validity in Interpretation*. New Haven: Yale Univ. Press, 1967.

——— *The Aims of Interpretation*. Chicago: Univ. of Chicago Press, 1978.

Hodgson, Brian Houghton. *Essays on the Languages, Literature, and Religion of Nepal and Tibet*. London: Williams & Norgate, 1874.

Hollinger, Robert. (ed.). *Hermeneutics and Praxis*. Notre Dame: Univ. of Notre Dame Press, 1985.

Howard, Roy J. *Three Faces of Hermeneutics*. Berkeley: Univ. of California Press, 1982.

Hoy, David Couzens. *The Critical Circle: Literature, History, and Philosophical Hermeneutics*. Berkeley: Univ. of California Press, 1978.

Inada, Kenneth K. and Nolan P. Jacobson, (eds.). *Buddhism and American Thinkers*. Albany: State University of New York Press, 1986.

Indian Council for Cultural Relations. *Indian Studies Abroad*. Bombay: Asia Publishing House, 1964.

Ingalls, D. H. H. *Materials for the Study of Navya-Nyāya Logic*. Cambridge: Harvard Univ. Press, 1951.

Jackson, Carl T. *The Oriental Religions and American Thought*. Westport, Conn.: Greenwood Press, 1981.

James, William. *Pragmatism*. New York: Longman, Green and Co., 1907.

Jayatilleke, K. N. *Early Buddhist Theory of Knowledge*. London: George Allen & Unwin, 1963.

Kabbani, Rana. *Europe's Myths of Orient*. Bloomington: Indiana Univ. Press, 1986.

Kalupahana, David J. *Nāgārjuna: The Philosophy of the Middle Way*. Albany: State University of New York Press, 1986.

Kant, Immanuel. *Immanuel Kant's Critique of Pure Reason*. Translated by Norman Kemp Smith. New York: St. Martin's Press, 1929.

Katz, Nathan, (ed.). *Buddhist and Western Philosophy*. New Delhi: Sterling Press, 1981.

——— "Nāgārjuna and Wittgenstein on Error." *Buddhist and Western Philosophy*. Edited by Nathan Katz. New Delhi: Sterling Press, 1981.

Keith, Arthur Berriedale. *Indian Logic and Atomism*. Oxford: Clarendon Press, 1921.

——— *Buddhist Philosophy in India and Ceylon*. Oxford: Clarendon Press, 1923.

Kenner, Hugh. *The Pound Era*. Berkeley: Univ. of California Press, 1971.

Kermode, Frank. *The Classic*. Cambridge: Harvard Univ. Press, 1983.

Kuhn, Thomas S. *The Structure of Scientific Revolutions*. Chicago: Univ. of Chicago Press, 1962.

——— *The Essential Tension*. Chicago: Univ. of Chicago Press, 1977.

Lach, Donald F. *Indian in the Eyes of Europe*. Chicago: Univ. of Chicago Press, 1965.

La Vallée Poussin, Louis de. *Bouddhisme: Opinions sur l'histoire de la dogmatique*. Paris: Gabriel Beauchesne et Cie., 1908.

Linge, David E. "Dilthey and Gadamer: Two Theories of Historical Understanding." *Journal of American Academy of Religion*, 41 (1973), pp. 536–553.

Louch, A. R. *Explanation and Human Action*. Berkeley: Univ. of California Press, 1966.

Lubac, Henri de. *La Recontre du bouddhisme et de l'occident*. Paris: Aubier, Editions Montaigne, 1952.

MacIntyre, Alasdair. *Against the Self-Images of the Age*. Notre Dame, Ind.: Univ. of Notre Dame Press, 1978.

Magliola, Robert. *Derrida on the Mend*. West Lafayette, Indiana: Purdue University Press, 1986.

Margolis, Joseph. *Pragmatism without Foundations*. Oxford: Basil Blackwell, 1986.

Matilal, Bimal Krishna. *The Navya-Nyāya Doctrine of Negation*. Cambridge: Harvard Univ. Press, 1968.

——— *Epistemology, Logic, and Grammar in Indian Philosophical Analysis*. The Hague: Mouton, 1971.

——— "A Critique of the Mādhyamika Position." *The Problem of Two Truths in Buddhism and Vedānta*. Edited by Mervyn Sprung. Dordrecht, Holland: D. Reidel, 1973.

——— *Nyāya-Vaiśeṣika*. Wiesbaden: Otto Harrassowitz, 1977.

Mehta, D. D. *Positive Sciences in the Vedas*. Delhi: Arnold Heineman Publishers, 1974.

Mehta, J. L. *India and the West*. Chico, Calif.: Scholars Press, 1985.

Mirsky, Jeanette. *Sir Aurel Stein*. Chicago: University of Chicago Press, 1977.

Mittal, K. K. *Materialism in Indian Thought*. Delhi: Munshiram Manoharlal, 1974.

Müller, Max. *The Six Systems of Indian Philosophy*. Varanasi: Chowkhamba Sanskrit Series Office, 1899.

Murti, T. R. V. *The Central Philosophy of Buddhism*. London: George Allen & Unwin, 1955.

——— "Saṃvṛti and Paramārtha." *The Problem of Two Truths in Buddhism and Vedānta*. Edited by Mervyn Sprung. Dordrecht, Holland: D. Reidel, 1974.

Nāgārjuna, *Mūlamadhyamikakārikā*. Translated by Kenneth K. Inada. Tokyo: Hokuseido Press, 1970.

——— *Vigrahavyāvartanī*. Translated by Kamaleswar Bhattacharya. Delhi: Motilal Banarsidass, 1978.

Newton-De Molina, David, (ed.). *On Literary Intention*. Edinburgh: Edinburgh Univ. Press, 1976.

Norris, Christopher. *The Deconstructive Turn*. London: Methuen & Co., 1983.

O'Neill, John, (ed.). *On Critical Theory*. New York: Seabury Press, 1976.

Palmer, Richard E., *Hermeneutics: Interpretation Theory in Schleiermacher, Dilthey, Heidegger, and Gadamer*. Evanston: Northwestern Univ. Press, 1969.

Pandeya, R. C. *The Problem of Meaning in Indian Philosophy*. Delhi: Motilal Banarsidass, 1963.

Pannikkar, K. M. *Asia and Western Dominance*. London: George Allen & Unwin, 1959.

Peiris, William. *The Western Contribution to Buddhism*. Delhi: Motilal Banarsidass, 1973.

Perl, Jeffrey M. *The Tradition of Return*. Princeton: Princeton Univ. Press, 1984.

——— *Skepticism and Modern Enmity: Before and After Eliot*. Baltimore: The Johns Hopkins Univ. Press, 1989.

Pilkington, A. E. *Bergson and His Influence: A Reassessment*. Cambridge: Cambridge Univ. Press, 1976.

Potter, Karl. "A Fresh Classification of India's Philosophical Systems." *Journal of Asian Studies*, 21 (1961), pp. 25–32.

——— *Presuppositions of India's Philosophies*. Englewood Cliffs, N.J.: Prentice-Hall, 1963.

——— (ed.). *Indian Metaphysics and Epistemology*. Princeton: Princeton Univ. Press, 1977.

Quine, Willard Van Orman. *From a Logical Point of View*. Cambridge: Harvard Univ. Press, 1953.

——— *Word and Object*. Cambridge: MIT Press, 1960.

Raja, C. Kunhan. *Some Fundamental Problems in Indian Philosophy*. New Delhi: Motilal Banarsidass, 1960.

Rajchman, John and Cornell West (eds.). *Post-Analytic Philosophy*. New York: Columbia Univ. Press, 1985.

Raju, P. T. *The Philosophical Traditions of India*. Pittsburgh: Univ. of Pittsburgh Press, 1971.

——— *Structural Depths of Indian Thought*. Albany: State Univ. of New York Press, 1985.

Riepe, Dale. *The Naturalistic Tradition in Indian Thought*. Seattle: Univ. of Washington Press, 1961.

——— *The Philosophy of India and Its Impact on American Thought*. Springfield, Ill.: Charles C Thomas, 1970.

Robinson, Richard H. "Some Logical Aspects of Nāgārjuna's System." *Philosophy East and West*. 6:4 (1957), pp. 291–308.

——— "Comments on 'Non-Western Studies: The Religious Approach' by Wilfred Cantwell Smith." *A Report on An Invitational Conference on the Study of Religion in the State University*. New Haven: The Society for Religion in Higher Education, nd. (Conference held October 23–25, 1964).

——— *Early Mādhyamika in India and China*. Madison: Univ. of Wisconsin Press, 1967.

——— "Classical Indian Philosophy." *Chapters in Indian Civilization*. Edited by Joseph W. Elder. Dubuque, Iowa: Kendall/Hunt, 1970.

Rorty, Richard (ed.). *The Linguistic Turn*. Chicago: Univ. of Chicago Press, 1967.

——— *Philosophy and the Mirror of Nature*. Princeton: Princeton Univ. Press, 1979.

——— *Consequences of Pragmatism*. Minneapolis: Univ. of Minnesota Press, 1982.

——— "Deconstruction and Circumvention." *Critical Inquiry*. 11:1 (1984), pp. 1–23.

——— (ed.). *Philosophy in History*. Cambridge: Cambridge Univ. Press, 1984.

Ruben, Walter. *Geschichte de Indichen Philosophie*. Berlin: Deutscher Verlag der Wissenschaften, 1954.

Said, Edward. *Orientalism*. New York: Random House, 1978.

——— *The World, the Text, and the Critic*. Cambridge: Harvard Univ. Press, 1983.

Saint-Hilaire, Jules Barthelemy. *Le Bouddha et sa religion*. Paris: Didier, Nouv. ed., 1862.

Sastri, H. Chatterjee. *The Philosophy of Nāgārjuna as Contained in the Ratnavali*. Calcutta: Saraswat Library, 1977.

Sedlar, Jean W. *India in the Mind of Germany*. Washington, D. C.: Univ. Press of America, 1982.

Scharfstein, Ben-Ami (ed.). *Philosophy East/Philosophy West*. New York: Oxford Univ. Press, 1978.

Schwab, Raymond. *The Oriental Renaissance*. New York: Columbia Univ. Press, 1984.

Schopenhauer, Arthur. *The World as Will and Representation,* 2 volumes. Translated by E. F. J. Payne. New York: Dover Press, 1966.

——— *Sämtliche Werke,* 6 volumes. Edited by Eduard Grisebach. Leipzig: Brockhaus, 1921–24.

Searle, John R. "Reiterating the Differences." *Glyph,* 1 (1977), pp. 198–208.

Sharma, Dhirendra. *The Differentiation Theory of Meaning in Indian Logic*. The Hague: Mouton, 1971.

Sharpe, Eric J. *Comparative Religion: A History*. La Salle, Illinois: Open Court, 1975.

Shastri, D. N. *Critique of Indian Realism*. Agra: Agra Univ. Press, 1964.

Shastri, D. R. *A Short History of Indian Materialism*. Calcutta: Univ. of Calcutta, 1930.

Shrivastava, S. N. L. *Śaṃkara and Bradley*. Delhi: Motilal Banarsidass, 1968.

Sinha, Jadunath. *Indian Realism*. London: Routledge & Kegan Paul, 1938.

——— *Indian Epistemology of Perception*. Calcutta: Sinha, 1969.

Skinner, B. F. *Science and Human Behavior*. New York: The Free Press, 1953.

Smart, Ninian. *Doctrine and Argument in Indian Philosophy*. London: George Allen & Unwin, 1964.

Sprung, Mervyn, ed. *The Problem of Two Truths in Buddhism and Vedānta*. Dordrecht, Holland: D. Reidel, 1974.

Stcherbatsky, Fyodor. *The Central Conception of Buddhism*. London: Royal Asiatic Society, 1923.

——— "Review of *Nirvāṇa* by Louis de La Vallée Poussin." *Bulletin of the School of Oriental Studies*, 4 (1926).

——— *Buddhist Logic*, 2 volumes. New York: Dover, 1962.

——— *The Conception of Buddhist Nirvaṇa*. Delhi: Motilal Banarsidass, 1968.

——— *Further Papers of Stcherbatsky*. Translated by Harish Chandra Gupta. Calcutta: Indian Studies, 1971.

Stout, Jeffrey. "What Is the Meaning of a Text?" *New Literary History*, 14 (1982–83), pp. 1–12.

——— *The Flight from Authority*. Notre Dame, Ind.: Univ. of Notre Dame Press, 1981.

Streng, Frederick. *Emptiness*. Nashville: Abingdon Press, 1967.

——— "The Significance of Pratītyasamutpāda for Understanding the Relationship between Samvṛti and Paramārthasatya in Nāgārjuna". *The Problem of Two Truths in Buddhism and Vedānta*. Dordrecht, Holland: D. Reidel, 1974.

Suzuki, D. T. *Outlines of Mahāyāna Buddhism*. New York: Schocken, 1963.

——— *On Indian Mahāyāna Buddhism*. New York: Harper Torchbooks, 1968.

Thomas, Edward J. *The History of Buddhist Thought*. London: Routledge & Kegan Paul, 1933.

Thurman, Robert. "Philosophical Nonegocentrism in Wittgenstein and Candrakīrti in Their Treatment of the Private Language Problem." *Philosophy East and West*, 30:3 (1980), pp. 321–337.

——— *Tsong Khapa's Speech of Gold in the Essence of True Eloquence*. Princeton: Princeton Univ. Press, 1984.

Tuck, Andrew P. and Jeffrey M. Perl. "The Hidden Advantage of Tradition: On the Significance of T. S. Eliot's Indic Studies." *Philosophy East and West*, 35:2 (1985), pp. 115–131.

Urquhart, W. S. *The Vedānta and Modern Thought*. Oxford: Oxford Univ. Press, 1928.

Vidyabhusana, Mahamahopadhyaya Satis Chandra. *A History of Indian Logic*. Delhi: Motilal Banarsidass, 1971.

Wach, Joachim. *Types of Religious Experience, Christian and Non-Christian*. Chicago: Univ. of Chicago Press, 1951.

Waldo, Ives. "Nāgārjuna and Analytic Philosophy." *Philosophy East and West*, 28:3 (1978), pp. 287–298.

Warder, A. K. *Indian Buddhism*. Delhi: Motilal Banarsidass, 1970.

——— *Outline of Indian Philosophy*. Delhi: Motilal Banarsidass, 1971.

Wayman, Alex. "Conze on Buddhism and European Parallels." *Philosophy East and West*, 13 (1963), 361–364.

——— "Who Understands the Four Alternatives of the Buddhist Texts?" *Philosophy East and West*, 27:1 (1977), pp. 3–21.

Welbon, Guy Richard. *The Buddhist Nirvāṇa and Its Western Interpreters*. Chicago: University of Chicago Press, 1968.

Williams, Monier. *Indian Wisdom*. Varanasi: Chowkhamba Sanskrit Series Office, 1963.

Willson, A. Leslie. *A Mythical Image: The Ideal of India in German Romanticism*. Durham, N. C.: Duke Univ. Press, 1964.

Winternitz, Maurice. *History of Indian Literature*, 2 volumes. New Delhi: Munshiram Manoharlal, 1927.

Wittgenstein, Ludwig. *Philosophical Investigations*. Translated by G. E. M. Anscombe. New York: Macmillan, 1953.

——— *Remarks on the Foundations of Mathematics*. New York: Macmillan, 1956.

——— *The Blue and Brown Books*. Oxford: Basil Blackwell, 1958.

——— *Zettel*. Translated by G. E. M. Anscombe. Oxford: Basil Blackwell, 1967.

——— *On Certainty*. Translated by Denis Paul and G. E. M. Anscombe. Oxford: Basil Blackwell, 1969.

Wright, Georg Henrik von. *Explanation and Understanding*. Ithaca, New York: Cornell Univ. Press, 1971.

INDEX